Graduate Texts in Mathematics 39

Springer
New York
Berlin
Heidelberg
Barcelona
Budapest
Hong Kong
London
Milan
Paris
Santa Clara
Singapore
Tokyo

Graduate Texts in Mathematics

1 Takeuti/Zaring. Introduction to Axiomatic Set Theory. 2nd ed.
2 Oxtoby. Measure and Category. 2nd ed.
3 Schaefer. Topological Vector Spaces.
4 Hilton/Stammbach. A Course in Homological Algebra. 2nd ed.
5 Mac Lane. Categories for the Working Mathematician.
6 Hughes/Piper. Projective Planes.
7 Serre. A Course in Arithmetic.
8 Takeuti/Zaring. Axiomatic Set Theory.
9 Humphreys. Introduction to Lie Algebras and Representation Theory.
10 Cohen. A Course in Simple Homotopy Theory.
11 Conway. Functions of One Complex Variable I. 2nd ed.
12 Beals. Advanced Mathematical Analysis.
13 Anderson/Fuller. Rings and Categories of Modules. 2nd ed.
14 Golubitsky/Guillemin. Stable Mappings and Their Singularities.
15 Berberian. Lectures in Functional Analysis and Operator Theory.
16 Winter. The Structure of Fields.
17 Rosenblatt. Random Processes. 2nd ed.
18 Halmos. Measure Theory.
19 Halmos. A Hilbert Space Problem Book. 2nd ed.
20 Husemoller. Fibre Bundles. 3rd ed.
21 Humphreys. Linear Algebraic Groups.
22 Barnes/Mack. An Algebraic Introduction to Mathematical Logic.
23 Greub. Linear Algebra. 4th ed.
24 Holmes. Geometric Functional Analysis and Its Applications.
25 Hewitt/Stromberg. Real and Abstract Analysis.
26 Manes. Algebraic Theories.
27 Kelley. General Topology.
28 Zariski/Samuel. Commutative Algebra. Vol.I.
29 Zariski/Samuel. Commutative Algebra. Vol.II.
30 Jacobson. Lectures in Abstract Algebra I. Basic Concepts.
31 Jacobson. Lectures in Abstract Algebra II. Linear Algebra.
32 Jacobson. Lectures in Abstract Algebra III. Theory of Fields and Galois Theory.
33 Hirsch. Differential Topology.
34 Spitzer. Principles of Random Walk. 2nd ed.
35 Alexander/Wermer. Several Complex Variables and Banach Algebras. 3rd ed.
36 Kelley/Namioka et al. Linear Topological Spaces.
37 Monk. Mathematical Logic.
38 Grauert/Fritzsche. Several Complex Variables.
39 Arveson. An Invitation to C^*-Algebras.
40 Kemeny/Snell/Knapp. Denumerable Markov Chains. 2nd ed.
41 Apostol. Modular Functions and Dirichlet Series in Number Theory. 2nd ed.
42 Serre. Linear Representations of Finite Groups.
43 Gillman/Jerison. Rings of Continuous Functions.
44 Kendig. Elementary Algebraic Geometry.
45 Loève. Probability Theory I. 4th ed.
46 Loève. Probability Theory II. 4th ed.
47 Moise. Geometric Topology in Dimensions 2 and 3.
48 Sachs/Wu. General Relativity for Mathematicians.
49 Gruenberg/Weir. Linear Geometry. 2nd ed.
50 Edwards. Fermat's Last Theorem.
51 Klingenberg. A Course in Differential Geometry.
52 Hartshorne. Algebraic Geometry.
53 Manin. A Course in Mathematical Logic.
54 Graver/Watkins. Combinatorics with Emphasis on the Theory of Graphs.
55 Brown/Pearcy. Introduction to Operator Theory I: Elements of Functional Analysis.
56 Massey. Algebraic Topology: An Introduction.
57 Crowell/Fox. Introduction to Knot Theory.
58 Koblitz. p-adic Numbers, p-adic Analysis, and Zeta-Functions. 2nd ed.
59 Lang. Cyclotomic Fields.
60 Arnold. Mathematical Methods in Classical Mechanics. 2nd ed.

continued after index

William Arveson

An Invitation to C^*-Algebras

Springer

William Arveson
Department of Mathematics
University of California
Berkeley, California 94720
USA

AMS Subject Classifications Primary: 46L05, 46L10, 46K10, 47C10
Secondary: 81A17, 81A54

Library of Congress Cataloging-in-Publication Data
Arveson, William.
An invitation to C*-algebras.
(Graduate texts in mathematics ; 39)
Bibliography.
Includes index.
1. C*-algebras. 2. Representations of algebras. 3. Hilbert space. I. Title. II. Series.
QA326.A78 512´.55 76-3656

Printed on acid-free paper.

Production managed by Allan Abrams; manufacturing supervised by Jeffrey Taub.
Printed and bound by Braun-Brumfield, Inc., Ann Arbor, MI.
Printed in the United States of America.

9 8 7 6 5 4 3 2 (Corrected second printing, 1998)

ISBN 0-387-90176-0 Springer-Verlag New York Berlin Heidelberg
ISBN 3-540-90176-0 Springer-Verlag Berlin Heidelberg New York SPIN 10671170

Preface

This book gives an introduction to C^*-algebras and their representations on Hilbert spaces. We have tried to present only what we believe are the most basic ideas, as simply and concretely as we could. So whenever it is convenient (and it usually is), Hilbert spaces become separable and C^*-algebras become GCR. This practice probably creates an impression that nothing of value is known about other C^*-algebras. Of course that is not true. But insofar as representations are concerned, we can point to the empirical fact that to this day no one has given a concrete parametric description of even the irreducible representations of any C^*-algebra which is not GCR. Indeed, there is metamathematical evidence which strongly suggests that no one ever will (see the discussion at the end of Section 3.4). Occasionally, when the idea behind the proof of a general theorem is exposed very clearly in a special case, we prove only the special case and relegate generalizations to the exercises.

In effect, we have systematically eschewed the Bourbaki tradition.

We have also tried to take into account the interests of a variety of readers. For example, the multiplicity theory for normal operators is contained in Sections 2.1 and 2.2. (it would be desirable but not necessary to include Section 1.1 as well), whereas someone interested in Borel structures could read Chapter 3 separately. Chapter 1 could be used as a bare-bones introduction to C^*-algebras. Sections 2.1 and 2.3 together contain the basic structure theory for type I von Neumann algebras, and are also largely independent of the rest of the book.

The level of exposition should be appropriate for a second year graduate student who is familiar with the basic results of functional analysis, measure theory, and Hilbert space. For example, we assume the reader knows the Hahn – Banach theorem, Alaoglu's theorem, the Krein – Milman theorem, the spectral theorem for normal operators, and the elementary theory of commutative Banach algebras. On the other hand, we have avoided making use of dimension theory and most of the more elaborate machinery of reduction theory (though we do use the notation for direct integrals in Sections 2.2 and 4.3). More regrettably, some topics have been left out merely to keep down the size of the book; for example, applications to the theory of unitary representations of locally compact groups are barely mentioned. To fill in these many gaps, the reader should consult the comprehensive monographs of Dixmier [6, 7].

A preliminary version of this manuscript was finished in 1971, and during the subsequent years was widely circulated in preprint form under the title *Representations of C^*-algebras*. The present book has been reorganized, and new material has been added to correct what we felt were serious omissions in the earlier version. It has been used as the basis for lectures in Berkeley and in Aarhus.

We are indebted to many colleagues and students who read the manuscript, pointed out errors, and offered constructive criticism. Special thanks go to Cecelia Bleecker, Larry Brown, Paul Chernoff, Ron Douglas, Dick Loebl, Donal O'Donovan, Joan Plastiras, and Erling Størmer.

This subject has more than its share of colorless and obscure terminology. In particular, one always has to choose between calling a C^*-algebra GCR, type I, or postliminal. The situation is no better in French: does postliminaire mean post-preliminary? In this book we have reverted to Kaplansky's original acronym, simply because it takes less space to write. More sensibly, we have made use of Halmos' symbol □ to signal the end of a proof.

Contents

An Invitation to C^*-Algebras

Fundamentals 1

This chapter contains what we consider to be the essentials of non-commutative C^*-algebra theory. This is the material that anyone who wants to work seriously with C^*-algebras needs to know. The most tractable C^*-algebras are those that can be related to compact operators in a certain specific way. These are the so-called GCR algebras, and they are introduced in Section 1.5, after a rather extensive discussion of C^*-algebras of compact operators in Section 1.4.

Representations are first encountered in Section 1.3; they remain near the center of discussion throughout the chapter, and indeed throughout the remainder of the book (excepting Chapter 3).

1.1. Operators and C^*-algebras

A *C^*-algebra of operators* is a subset $\mathscr{A}$ of the algebra $\mathscr{L}(\mathscr{H})$ of all bounded operators on a Hilbert space $\mathscr{H}$, which is closed under all of the algebraic operations on $\mathscr{L}(\mathscr{H})$ (addition, multiplication, multiplication by complex scalars), is closed in the norm topology of $\mathscr{L}(\mathscr{H})$, and most importantly is closed under the adjoint operation $T \mapsto T^*$ in $\mathscr{L}(\mathscr{H})$. Every operator T on $\mathscr{H}$ determines a C^*-algebra $C^*(T)$, namely the smallest C^*-algebra containing both T and the identity. It is more or less evident that $C^*(T)$ is the norm closure of all polynomials $p(T, T^*)$, where $p(x, y)$ ranges over all polynomials in the two free (i.e., noncommuting) variables x and y having complex coefficients. However since T and T^* do not generally commute, these polynomials in T and T^* are of little use in answering questions, and in particular the above remark sheds no light on the structure of $C^*(T)$. Nevertheless, $C^*(T)$ contains much information about T, and one could view this book as a description of what that information is and how one goes about extracting it.

We will say that two operators S and T (acting perhaps, on different Hilbert spaces) are *algebraically equivalent* if there is a $*$-isomorphism (that is, an isometric $*$-preserving isomorphism) of $C^*(S)$ onto $C^*(T)$ which carries S into T. Note that this is more stringent than simply requiring that $C^*(S)$ and $C^*(T)$ be $*$-isomorphic. We will see presently that two normal operators are algebraically equivalent if and only if they have the same spectrum; thus one may think of algebraically equivalent nonnormal operators as having the same "spectrum" in some generalized sense, which will be made more precise in Chapter 4.

We now collect a few generalities. A (general) *C^*-algebra* is a Banach algebra A having an involution $*$ (that is, a conjugate-linear map of A into itself satisfying $x^{**} = x$ and $(xy)^* = y^*x^*$, $x, y \in A$) which satisfies $\|x^*x\| = \|x\|^2$ for all $x \in A$. It is very easy to see that a C^*-algebra of operators on a Hilbert space is a C^*-algebra, and we will eventually prove a theorem of Gelfand and Naimark which asserts the converse: every C^*-algebra is isometrically $*$-isomorphic with a C^*-algebra of operators on a Hilbert space (Theorem 1.7.3).

Let A be a *commutative* C^*-algebra. Then in particular A is a commutative Banach algebra, and therefore the set of all nonzero complex homomorphisms of A is a locally compact Hausdorff space in its usual topology. This space will be called the *spectrum* of A, and it is written $\hat{A}$. A standard result asserts that $\hat{A}$ is compact iff A contains a multiplicative identity. Now the Gelfand map is generally a homomorphism of A into the Banach algebra $C(\hat{A})$ of all continuous complex valued functions on $\hat{A}$ vanishing at ∞. In this case, however, much more is true.

Theorem 1.1.1. *The Gelfand map is an isometric $*$-isomorphism of A onto $C(\hat{A})$.*

Here, the term $*$-isomorphism means that, in addition to the usual properties of an isomorphism, $x^* \in A$ gets mapped into the complex conjugate of the image of x. We will give the proof of this theorem for the case where A contains an identity 1; the general case follows readily from this by the process of adjoining an identity (Exercise 1.1.H).

First, let $\omega \in \hat{A}$. Then we claim $\omega(x^*) = \overline{\omega(x)}$ for all $x \in A$. This reduces to proving that $\omega(x)$ is real for all $x = x^*$ in A (since every $x \in A$ can be written $x = x_1 + ix_2$, with $x_i = x_i^* \in A$). Therefore choose $x = x^* \in A$, and for every real number t define $u_t = e^{itx}$ (for any element $z \in A$, e^z is defined by the convergent power series $\sum_{n=0}^{\infty} z^n/n!$, and the usual manipulations show that $e^{z+w} = e^z e^w$ since z and w commute). By examining the power series we see that $u_t^* = e^{-itx}$, and hence $u_t^* u_t = e^{-itx+itx} = 1$. Thus $\|u_t\|^2 = \|u_t^* u_t\| = \|1\| = 1$, and since the complex homomorphism ω has norm 1 we conclude $\exp t\, \mathscr{Re}\, i\omega(x) = |e^{\omega(itx)}| = |\omega(u_t)| \leq 1$, for all $t \in \mathbb{R}$. This can only mean $\mathscr{Re}\, i\omega(x) = 0$, and hence $\omega(x)$ is real.

Now let $\gamma(x)$ denote the image of x in $C(\hat{A})$, i.e., $\gamma(x)(\omega) = \omega(x)$, $\omega \in \hat{A}$. Then we have just proved $\gamma(x^*) = \overline{\gamma(x)}$, and we now claim $\|\gamma(x)\| = \|x\|$. Indeed the left side is the spectral radius of x which, by the Gelfand-Mazur theorem, is $\lim_n \|x^n\|^{1/n}$. But if $x = x^*$ then we have $\|x\|^2 = \|x^*x\| = \|x^2\|$; replacing x with x^2 gives $\|x\|^4 = \|x^2\|^2 = \|x^4\|$, and so on inductively, giving $\|x\|^{2^n} = \|x^{2^n}\|$, $n \geqslant 1$. This proves $\|x\| = \lim_n \|x^n\|^{1/n}$ if $x = x^*$, and the case of general x reduces to this by the trick $\|\gamma(x)\|^2 = \|\overline{\gamma(x)}\gamma(x)\| = \|\gamma(x^*x)\| = \|x^*x\| = \|x\|^2$, applying the above to the self-adjoint element x^*x.

Thus γ is an isometric $*$-isomorphism of A onto a closed self-adjoint subalgebra of $C(\hat{A})$ containing 1; since $\gamma(A)$ always separates points, the proof is completed by an application of the Stone–Weierstrass theorem. □

An element x of a C^*-algebra is called *normal* if $x^*x = xx^*$. Note that this is equivalent to saying that the sub C^*-algebra generated by x is commutative.

Corollary. *If x is a normal element of a C^*-algebra with identity, then the norm of x equals its spectral radius.*

PROOF. Consider x to be an element of the commutative C^*-algebra it generates (together with the identity). Then the assertion follows from the fact that the Gelfand map is an isometry. □

Theorem 1.1.1 is sometimes called the *abstract spectral theorem*, since it provides the basis for a powerful functional calculus in C^*-algebras. In order to discuss this, let us first recall that if B is a Banach subalgebra of a Banach algebra A with identity 1, such that $1 \in B$, then an element x in B has a spectrum $\mathrm{sp}_A(x)$ relative to A as well as a spectrum $\mathrm{sp}_B(x)$ relative to B, and in general one has $\mathrm{sp}_A(x) \subseteq \mathrm{sp}_B(x)$. Of course, the inclusion is often proper. But if A is a C^*-algebra and B is a C^*-subalgebra, then the two spectra must be the same. To indicate why this is so, we will show that if $x \in B$ is invertible in A, then x^{-1} belongs to B (a moment's thought shows that the assertion reduces to this). For that, note that x^* is invertible, and since the element $(x^*x)^{-1}x^*$ is clearly a *left* inverse for x, we must have $x^{-1} = (x^*x)^{-1}x^*$. So to prove that $x^{-1} \in B$, it suffices to show that $(x^*x)^{-1} \in B$. Actually, we will show that x^*x is invertible in the still smaller C^*-algebra B_0 generated by x^*x and e. For since B_0 is commutative, 1.1.1 implies that the spectrum (relative to B_0) of the self-adjoint element x^*x is real, and in particular this relative spectrum is its own boundary, considered as a subset of the complex plane. By the spectral permanence theorem ([23], p. 33), the latter coincides with $\mathrm{sp}_A(x^*x)$. Because $0 \notin \mathrm{sp}_A(x^*x)$, we conclude that x^*x is invertible in B_0.

These remarks show in particular that it is unambiguous to speak of the spectrum of an operator T on a Hilbert space $\mathscr{H}$, so long as it is taken relative to a C^*-algebra. Thus, the spectrum of T in the traditional sense (i.e., relative to $\mathscr{L}(\mathscr{H})$) is the same as the spectrum of T relative to the subalgebra $C^*(T)$. They also show that the spectrum of a self-adjoint element of an arbitrary C^*-algebra (commutative or not) is always real.

We can now deduce the functional calculus for normal elements of C^*-algebras. Fix such an element x in a C^*-algebra with identity, and let B be the C^*-algebra generated by x and e. Define a map of $\hat{B}$ into $\mathbb{C}$ as follows: $\omega \to \omega(x)$. This is continuous and 1—1, thus since $\hat{B}$ is compact it is a homeomorphism of $\hat{B}$ onto its range. By the preceding discussion the range of this map is $\mathrm{sp}_A(x) = \mathrm{sp}(x)$. So this map induces, by composition, an isometric $*$-isomorphism of $C(\mathrm{sp}(x))$ onto B. It is customary to write the image of a function $f \in C(\mathrm{sp}(x))$ under this isomorphism as $f(x)$. Note that the formula suggested by this notation reduces to the expected thing when f is a polynomial in ζ and $\bar{\zeta}$; for example, if $f(\zeta) = \zeta^2\bar{\zeta}$ then $f(x) = x^2x^*$. This process of "applying" continuous functions on $\mathrm{sp}(x)$ to x is called the *functional calculus.*

In particular, when T is a normal operator on a Hilbert space we have defined expressions of the form $f(T)$, $f \in C(\mathrm{sp}(T))$. In this concrete setting one can even extend the functional calculus to arbitrary bounded (or even unbounded) Borel functions defined on $\mathrm{sp}(T)$, but we shall have no particular need for that in this book. It is now a simple matter to prove:

Theorem 1.1.2. *Let S and T be normal operators. Then S and T are algebraically equivalent if, and only if, they have the same spectrum.*

Proof. Assume first that $\mathrm{sp}(S) = \mathrm{sp}(T)$. Then by the above we have $\|f(S)\| = \sup\{|f(z)| : z \in \mathrm{sp}(S)\} = \|f(T)\|$, for every continuous function f on $\mathrm{sp}(S)$. This shows that the map $\phi : f(S) \to f(T)$, $f \in C(\mathrm{sp}(S))$, is an isometric $*$-isomorphism of $C^*(S)$ on $C^*(T)$ which carries S to T. Conversely, if such a ϕ exists, then the spectrum of S relative to $C^*(S)$ must equal the spectrum of $\phi(S) = T$ relative to $C^*(T)$. By the preceding remarks, this implies $\mathrm{sp}(S) = \mathrm{sp}(T)$. □

Exercises

1.1.A. Let e be an element of a C^*-algebra which satisfies $ex = x$ for every $x \in A$. Show that e is a unit, $e = e^*$, and $\|e\| = 1$.

1.1.B. Let A be a Banach algebra having an involution $x \to x^*$ which satisfies $\|x\|^2 \leqslant \|x^*x\|$ for every x. Show that A is a C^*-algebra.

1.1.C. (Mapping theorem.) Let x be a self-adjoint element of a C^*-algebra with unit and let $f \in C(\mathrm{sp}(x))$. Show that the spectrum of $f(x)$ is $f(\mathrm{sp}(x))$.

1.1.D. Let A be the algebra of all continuous complex-valued functions, defined on the closed disc $D = \{|z| \leqslant 1\}$ in the complex plane, which are analytic in the interior of D.

a. Show that A is a commutative Banach algebra with unit, relative to the norm $\|f\| = \sup_{|z| \leqslant 1} |f(z)|$.

b. Show that $f^*(z) = \overline{f(\bar{z})}$ defines an isometric involution in A.

c. Show that not every complex homomorphism ω of A satisfies $\omega(f^*) = \overline{\omega(f)}$.

* Could we have x normal?

1.1.E. Let A be a C^*-algebra without unit. Show that, for every x in A:

$$\|x\| = \sup_{\|y\| \leqslant 1} \|xy\|.$$

1.1.F. Let S and T be normal operators on Hilbert spaces $\mathscr{H}$ and $\mathscr{K}$. Show that $C^*(S)$ is $*$-isomorphic to $C^*(T)$ iff $\operatorname{sp}(S)$ is homeomorphic to $\operatorname{sp}(T)$.

1.1.G. Let $f:\mathbb{R} \to \mathbb{C}$ be a continuous function and let A be a C^*-algebra with unit. Show that the mapping $x \mapsto f(x)$ is a continuous function from $\{x \in A : x = x^*\}$ into A.

1.1.H. (Exercise on adjoining a unit.) Let A be a C^*-algebra without unit, and for each x in A let L_x be the linear operator on A defined by $y \mapsto xy$. Let B be the set of all operators on A of the form $\lambda 1 + L_x$, $\lambda \in \mathbb{C}$, $x \in A$.

Show that B is a C^*-algebra with unit relative to the operator norm and the involution $(\lambda 1 + L_x)^* = \bar{\lambda} 1 + L_{x^*}$, and that $x \mapsto L_x$ is an isometric $*$-isomorphism of A onto a closed ideal in B of codimension 1. [*Hint*: use 1.1.B.]

1.1.I. Discuss briefly how the functional calculus (for self-adjoint elements) must be modified for C^*-algebras with no unit. In particular, explain why $\sin x$ makes sense for every self-adjoint element x but $\cos x$ does not. [*Hint*: use 1.1.H to define the spectrum of an element in a non-unital C^*-algebra.]

1.2. Two Density Theorems

There are two technical results which are extremely useful in dealing with $*$-algebras of operators. We will discuss these theorems in this section and draw out a few applications.

The *null space* of a set $\mathscr{S} \subseteq \mathscr{L}(\mathscr{H})$ of operators is the closed subspace of all vectors $\xi \in \mathscr{H}$ such that $S\xi = 0$ for all $S \in \mathscr{S}$. The *commutant* of $\mathscr{S}$ (written $\mathscr{S}'$) is the family of operators which commute with each element of $\mathscr{S}$. Note that $\mathscr{S}'$ is always closed under the algebraic operations, contains the identity operator, and is closed in the weak operator topology. Moreover, if $\mathscr{S}$ is *self-adjoint*, that is $\mathscr{S} = \mathscr{S}^*$ is closed under the $*$-operation, then so is $\mathscr{S}'$. Now it is evident that $\mathscr{S}$ is always contained in $\mathscr{S}''$, but even when $\mathscr{S}$ is a weakly closed algebra containing the identity the inclusion may be proper. According to the following celebrated theorem of von Neumann, however, one has $\mathscr{S} = \mathscr{S}''$ if in addition $\mathscr{S}$ is self-adjoint.

Theorem 1.2.1 Double commutant theorem. *Let $\mathscr{A}$ be a self-adjoint algebra of operators which has trivial null space. Then $\mathscr{A}$ is dense in $\mathscr{A}''$ in both the strong and the weak operator topologies.*

Proof. Let $\mathscr{A}_w$ and $\mathscr{A}_s$ denote the weak and strong closures of $\mathscr{A}$. Then clearly $\mathscr{A}_s \subseteq \mathscr{A}_w \subseteq \mathscr{A}''$, and it suffices to show that each operator $T \in \mathscr{A}''$ can be strongly approximated by operators in $\mathscr{A}$; that is, for every $\varepsilon > 0$, every $n = 1, 2, \ldots$, and every choice of n vectors $\xi_1, \xi_2, \ldots, \xi_n \in \mathscr{H}$, there is an operator $S \in \mathscr{A}$ such that $\sum_{k=1}^{n} \|T\xi_k - S\xi_k\|^2 < \varepsilon^2$.

Consider first the case $n = 1$, and let P be the projection onto the closed subspace $[\mathscr{A}\xi_1]$. Note first that P commutes with $\mathscr{A}$. Indeed the range of P is invariant under $\mathscr{A}$; since $\mathscr{A} = \mathscr{A}^*$, so is the range of $P^\perp = I - P$, and this implies $P \in \mathscr{A}'$. Next observe that $\xi_1 \in [\mathscr{A}\xi_1]$, or equivalently, $P^\perp\xi_1 = 0$. For if $S \in \mathscr{A}$ then $SP^\perp\xi_1 = P^\perp S\xi_1 = 0$ (because $S\xi_1 \in [\mathscr{A}\xi_1]$ and $P^\perp$ is zero on $[\mathscr{A}\xi_1]$). Since $\mathscr{A}$ has trivial null space we conclude $P^\perp\xi_1 = 0$. Finally, since T must commute with $P \in \mathscr{A}'$ it must leave the range of P invariant, and thus $T\xi_1 \in$ range $P = [\mathscr{A}\xi_1]$. This means we can find $S \in \mathscr{A}$ such that $\|T\xi_1 - S\xi_1\| < \varepsilon$, as required.

Now the case of general $n \geqslant 2$ is reduced to the above by the following device. Fix n, and let $\mathscr{H}_n = \mathscr{H} \oplus \cdots \oplus \mathscr{H}$ be the direct sum of n copies of the underlying Hilbert space $\mathscr{H}$. Choose $\xi_1, \ldots, \xi_n \in \mathscr{H}$ and define $\eta \in \mathscr{H}_n$ by $\eta = \xi_1 \oplus \xi_2 \oplus \cdots \oplus \xi_n$. Let $\mathscr{A}_n \subseteq \mathscr{L}(\mathscr{H}_n)$ be the $*$-algebra of all operators of the form $\{S \oplus S \oplus \cdots \oplus S : S \in \mathscr{A}\}$. Thus each element of $\mathscr{A}_n$ can be expressed as a diagonal $n \times n$ operator matrix

$$\begin{pmatrix} S & & & 0 \\ & S & & \\ & & \ddots & \\ 0 & & & S \end{pmatrix}$$

$S \in \mathscr{A}$. The reader can see by a straightforward calculation that an $n \times n$ operator matrix (T_{ij}), $T_{ij} \in \mathscr{L}(\mathscr{H})$, commutes with $\mathscr{A}_n$ iff each entry T_{ij} belongs to $\mathscr{A}'$. This gives a representation for $\mathscr{A}_n'$ as operator matrices, and now a similar calculation shows that (T_{ij}) commutes with $\mathscr{A}_n'$ iff (T_{ij}) has the form

$$\begin{pmatrix} T & & & 0 \\ & T & & \\ & & \ddots & \\ 0 & & & T \end{pmatrix}$$

with $T \in \mathscr{A}''$. Thus we have a representation for $\mathscr{A}_n''$. Now choose $T \in \mathscr{A}''$ and let $T_n = T \oplus T \oplus \cdots \oplus T$. Then $T_n \in \mathscr{A}_n''$ so that the argument already given shows that $T_n\eta \in [\mathscr{A}_n\eta]$, thus we can find $S \in \mathscr{A}$ such that $S_n\eta$ is within ε of $T_n\eta$ in the norm of $\mathscr{H}_n$. In other words, $\sum_{k=1}^n \|T\xi_k - S\xi_k\|^2 < \varepsilon^2$, as required. □

By definition, a *von Neumann algebra* is a self-adjoint subalgebra $\mathscr{R}$ of $\mathscr{L}(\mathscr{H})$ which contains the identity and is closed in the weak operator topology. Note that 1.2.1. asserts that such an $\mathscr{R}$ satisfies $\mathscr{R} = \mathscr{R}''$, and this gives a convenient criterion for an operator T to belong to $\mathscr{R}$: one simply checks to see if T commutes with $\mathscr{R}'$. As an illustration of this, let us consider the polar decomposition. That is, let $T \in \mathscr{L}(\mathscr{H})$, and let $|T|$ denote the positive square root of the positive operator T^*T (via the functional calculus). Then $|T| \in C^*(T^*T)$, and in particular $|T|$ belongs to the von Neumann algebra generated by T. We want to define a certain operator U such that $U|T| = T$. Note first that for all $\xi = \mathscr{H}$ we have $\||T|\xi\|^2 = (|T|\xi, |T|\xi) = (|T|^2\xi, \xi) = (T^*T\xi, \xi) = (T\xi, T\xi) = \|T\xi\|^2$. Therefore the map $U:|T|\xi \to T\xi$, $\xi \in \mathscr{H}$, extends uniquely to a linear isometry of the closed range of $|T|$ onto the closed range of T. Extend U to a bounded operator on $\mathscr{H}$ by putting $U = 0$ on the orthogonal complement of $|T|\mathscr{H}$. Then U is a partial isometry (i.e., U^*U is a projection) whose initial space is $[|T|\mathscr{H}]$ and which satisfies $U|T| = T$. It is easy to see that these properties determine U uniquely, and the above formula relating U and $|T|$ to T is called the *polar decomposition* of T. Now we want to show that U belongs to the von Neumann algebra generated by T. By 1.2.1, it suffices to show that U commutes with every operator Z which commutes with both T and T^*. Now in particular Z commutes with the self-adjoint operator $|T|$, and therefore Z leaves both $|T|\mathscr{H}$ and $(|T|\mathscr{H})^\perp$ invariant. In particular Z leaves the null space of $U(=(|T|\mathscr{H})^\perp)$ invariant and so $ZU = UZ = 0$ on the null space of U. Thus it suffices to show that $ZU = UZ$ on every vector of the form $|T|\xi$, $\xi \in \mathscr{H}$. But $ZU|T|\xi = ZT\xi = TZ\xi$, while $UZ|T|\xi = U|T|Z\xi = TZ\xi$, and we are done. This proves the following

Corollary. *Let $T = U|T|$ be the polar decomposition of an operator $T \in \mathscr{L}(\mathscr{H})$. Then both factors U and $|T|$ belong to the von Neumann algebra generated by T.*

The following density theorem is a special case of a theorem of Kaplansky [16]. For a set of operators $\mathscr{S}$ we will write ball $\mathscr{S}$ for the closed unit ball in $\mathscr{S}$, ball $\mathscr{S} = \{S \in \mathscr{S}: \|S\| \leqslant 1\}$.

Theorem 1.2.2. *Let $\mathscr{A}$ be a self-adjoint algebra of operators and let $\mathscr{A}_s$ be the closure of $\mathscr{A}$ in the strong operator topology. Then every self-adjoint element in* ball $\mathscr{A}_s$ *can be strongly approximated by self-adjoint elements in* ball $\mathscr{A}$.

Proof. Note first that every self-adjoint element in the unit ball of the *norm* closure of $\mathscr{A}$ can be norm-approximated by self-adjoint elements in ball $\mathscr{A}$. Thus we can assume $\mathscr{A}$ is norm closed.

Now since the $*$-operation is not strongly continuous, we cannot immediately assert that the strong closure of the convex set $\mathscr{S}$ of self-adjoint elements of $\mathscr{A}$ contains $\{T \in \mathscr{A}_s: T = T^*\}$. But its *weak* closure does (because if a net S_n converges to $T = T^*$ strongly, then the real parts of S_n converge weakly to T), and moreover since the weak and strong operator topologies have the same continuous linear functionals (Exercise 1.2.E) they must also have the same closed convex sets. Thus we see in this way that the strong closure of $\mathscr{S}$ contains the self-adjoint elements of $\mathscr{A}_s$.

Now consider the continuous functions $f:\mathbb{R} \to [-1, +1]$ and $g:[-1, +1] \to \mathbb{R}$ defined by $f(x) = 2x(1 + x^2)^{-1}$ and $g(y) = y(1 + \sqrt{1 - y^2})^{-1}$. Then we have $f \circ g(y) = y$, for all $y \in [-1, +1]$, and clearly $|f(x)| \leqslant 1$ for all $x \in \mathbb{R}$. We claim that the map $S \mapsto f(S)$ is strongly continuous on the set of all self-adjoint operators on $\mathscr{H}$. Granting that for a moment, note that 1.2.2 follows. For if $T = T^* \in \mathscr{A}_s$ is such that $\|T\| \leqslant 1$, then $S_0 = g(T)$ is a self-adjoint element of $\mathscr{A}_s$, so that by the preceding paragraph there is a net S_n of self-adjoint elements of $\mathscr{A}$ which converges strongly to S_0. Hence $f(S_n) \to f(S_0)$ strongly. Now $f(S_n)$ is self-adjoint, belongs to $\mathscr{A}$ (because $\mathscr{A}$ is norm closed), and has norm $\leqslant 1$ since $|f| \leqslant 1$ on $\mathbb{R}$. On the other hand, $f \circ g(y) = y$ on $[-1, 1]$ implies $f(S_0) = f \circ g(T) = T$, because $\|T\| \leqslant 1$, and this proves T is the strong limit of self-adjoint elements of ball $\mathscr{A}$.

Finally, the fact that $f(S) = 2S(I + S^2)^{-1}$ is strongly continuous follows after a moments reflection upon the operator identity

$$2[f(S) - f(S_0)] = 4(1 + S^2)^{-1}(S - S_0)(1 + S_0^2)^{-1} - f(S)(S - S_0)f(S_0),$$

considering S tending strongly to S_0 □

Kaplansky also proved that ball $\mathscr{A}$ is strongly dense in ball $\mathscr{A}_s$. That is not obvious from what we have said, but a simple trick using 2×2 operator matrices allows one to deduce that from 1.2.2 (Exercise 1.2.D).

Corollary. *Let $\mathscr{A}$ be a self-adjoint algebra of operators on a separable Hilbert space $\mathscr{H}$. Then for every operator T in the strong closure of $\mathscr{A}$, there is a sequence $T_n \in \mathscr{A}$ such that $T_n \to T$ in the strong operator topology.*

Proof. We can assume $\|T\| \leqslant 1$, and since we can argue separately with the real and imaginary parts of T, we can assume $T = T^*$. Let $\xi_1, \xi_2, \ldots$ be a countable dense set in $\mathscr{H}$. By 1.2.2, for each $n \geqslant 1$, we can find a self-adjoint element T_n in $\mathscr{A}$ such that $\|T_n\| \leqslant 1$ and $\|T_n\xi_k - T\xi_k\| < 1/n$ for $k = 1, 2, \ldots, n$. Thus $T_n \to T$ strongly on the dense set $\{\xi_1, \xi_2, \ldots\}$ of $\mathscr{H}$, and since $\|T_n\| \leqslant 1$, the corollary follows. □

This corollary shows that in the separable case, the strong closure of a C^*-algebra of operators can be achieved by adjoining to the algebra all limits of its strongly convergent sequences.

A C^*-algebra is *separable* if it has countable norm-dense subset. A separable C^*-algebra is obviously countably generated (a countable dense set clearly generates), and the reader can easily verify the converse: every countably generated C^*-algebra is separable. We conclude this section by pointing out a useful relation between separably-acting von Neumann algebras and separable C^*-algebras.

Let $\mathscr{H}$ be a Hilbert space. Then it is well known that the closed unit ball in $\mathscr{L}(\mathscr{H})$ is compact in the relative weak operator topology ([7], p. 34). Moreover, note that if $\mathscr{H}$ is separable then ball $\mathscr{L}(\mathscr{H})$ is a compact *metric space*. Indeed, if $u_1, u_2, \ldots$ is a countable dense set of unit vectors in $\mathscr{H}$ then the function

$$d(S, T) = \sum_{i,j=1}^{\infty} 2^{-i-j}|(Su_i - Tu_i, u_j)|$$

defines a metric on ball $\mathscr{L}(\mathscr{H}) \times$ ball $\mathscr{L}(\mathscr{H})$ which, as is easily seen, gives rise to the weak operator topology on ball $\mathscr{L}(\mathscr{H})$. Thus we see in particular that ball $\mathscr{L}(\mathscr{H})$ is a separable complete metric space.

Now if $\mathscr{R}$ is any von Neumann algebra acting on a separable Hilbert space then ball $\mathscr{R}$ is a weakly closed subset of ball $\mathscr{L}(\mathscr{H})$ and therefore has a countable weakly dense subset $T_1, T_2, \ldots$. Thus the C^*-algebra generated by $\{T_1, T_2, \ldots\}$ is a separable C^*-algebra contained in $\mathscr{R}$ whose weak closure coincides with $\mathscr{R}$, and we deduce the following result.

Proposition 1.2.3. *For every von Neumann algebra $\mathscr{R}$ acting on a separable Hilbert space there is a separable C^*-algebra $\mathscr{A} \subset \mathscr{R}$ which is weakly dense in $\mathscr{R}$.*

Exercises

1.2.A. An operator $A \in \mathscr{L}(\mathscr{H})$ is said to be *bounded below* if there is an $\varepsilon > 0$ such that $\|Ax\| \geqslant \varepsilon\|x\|$ for every $x \in \mathscr{H}$. Prove that for such an operator A, $C^*(A)$ contains both factors of the polar decomposition of A.

1.2.B. (Exercise on the polar decomposition.) Let $T \in \mathscr{L}(\mathscr{H})$ have the polar decomposition $T = U|T|$.

a. Show that if R is a positive operator on $\mathscr{H}$ and V is a partial isometry whose initial space is $[R\mathscr{H}]$, and which satisfy $T = VR$, then $V = U$ and $R = |T|$.

b. Show that U maps the subspace $\ker T^\perp$ onto $[T\mathscr{H}]$.

1.2.C. Let $\mathscr{R}$ be a von Neumann algebra acting on $\mathscr{H}$, let $\mathscr{M}$ be a subspace of $\mathscr{H}$ whose projection P belongs to $\mathscr{R}$, and let $T \in \mathscr{R}$.

a. Show that the projection Q on $[T\mathscr{M}]$ belongs to $\mathscr{R}$.

b. Show that there is a partial isometry $U \in \mathscr{R}$ satisfying $U^*U = P$ and $UU^* = Q$. (This shows that, in any von Neumann algebra $\mathscr{R}$, the partial isometries do as good a job of moving subspaces around as arbitrary operators in $\mathscr{R}$.)

1.2.D. (Exercise on Kaplansky's density theorem.) Let $\mathscr{R}$ be a von Neumann algebra on $\mathscr{H}$.

a. Show that the set of all operators on $\mathscr{H} \oplus \mathscr{H}$ which admit a 2×2 operator matrix representation of the form

$$\begin{pmatrix} A & B \\ C & D \end{pmatrix},$$

with A, B, C, D in $\mathscr{R}$, is a von Neumann algebra on $\mathscr{H} \oplus \mathscr{H}$.

b. Let A be a $*$-algebra of operators on $\mathscr{H}$, and let T be an operator in the unit ball of the strong closure of $\mathscr{A}$. Use 1.2.2 to show that T can be strongly approximated by operators in the unit ball of $\mathscr{A}$. [*Hint*: consider the operator

$$\begin{pmatrix} 0 & T \\ T^* & 0 \end{pmatrix}$$

on $\mathscr{H} \oplus \mathscr{H}$.]

1.2.E. (Exercise on the weak and strong operator topologies.) Let $\mathscr{H}$ be a Hilbert space and let f be a linear functional on $\mathscr{L}(\mathscr{H})$ which is continuous in the strong operator topology.

a. Show that there exist vectors $\xi_1, \ldots, \xi_n \in \mathscr{H}$ such that

$$|f(T)| \leq (\|T\xi_1\|^2 + \cdots + \|T\xi_n\|^2)^{1/2}, \qquad T \in \mathscr{L}(\mathscr{H}).$$

b. With $\xi_1, \ldots, \xi_n$ as in part a, show that there exist vectors $\eta_1, \ldots, \eta_n$ in $\mathscr{H}$ such that f has a representation

$$f(T) = (T\xi_1, \eta_1) + \cdots + (T\xi_n, \eta_n),$$

$T \in \mathscr{L}(\mathscr{H})$. [*Hint*: apply the Riesz lemma to a certain linear functional on the Hilbert sum of n copies of $\mathscr{H}$.]

c. Deduce that the weak and strong operator topologies have the same continuous linear functionals and the same closed convex sets.

1.3. Ideals, Quotients, and Representations

In this section we will discuss a few basic properties of C^*-algebras and introduce some terminology. By an *ideal* in a C^*-algebra we will always mean a *closed* two sided ideal. It is sometimes useful to consider left or right ideals, but we shall not have to do so here.

Many C^*-algebras do not have identities. This is particularly true of ideals in a given C^*-algebra, considered as C^*-algebras in their own right. Very often this lack is merely an annoyance, and the difficulty it presents can be circumvented by simply adjoining an identity. For instance, the spectrum of an element of a C^*-algebra is a concept which obviously requires a unit. The most direct way of defining the spectrum of an element x of a non-unital C^*-algebra A is to consider x to be an element of the C^*-algebra A_1 obtained from A by adjoining a unit, where $\operatorname{sp}(x)$ has an obvious meaning.

But frequently, and especially when dealing with ideals, it is necessary to make use of a more powerful device. An *approximate identity* (or *approximate unit*) for a Banach algebra A is a net $\{e_\lambda\}$ of elements of A satisfying

(i) $\|e_\lambda\| = 1$, *for every* λ
(ii) $\lim_\lambda \|xe_\lambda - x\| = \lim_\lambda \|e_\lambda x - x\| = 0$, *for every* $x \in A$.

When A is a C^*-algebra, one usually makes additional requirements of the net $\{e_\lambda\}$ (see Section 1.8). It is a basic property of C^*-algebras, as opposed to more general Banach algebras, that approximate identities always exist. We will eventually prove that assertion (cf. 1.8.2); but for our purposes in this section, all we shall require is a simple result which implies that one-sided approximate units exist "locally."

Proposition 1.3.1. *Let A be a C^*-algebra and let J be an ideal in A. Then for every x in J there is a sequence $e_1, e_2, \ldots$ of self-adjoint elements of J satisfying*

(i) $\operatorname{sp}(e_n) \subseteq [0, 1]$ *for every* n;
(ii) $\lim\|xe_n - x\| = 0$.

Proof. Consider first the case where A has an identity e and $x = x^*$. Define $e_n \in A$ (via the functional calculus) by

$$e_n = nx^2(e + nx^2)^{-1}.$$

The function $f_n:\mathbb{R} \to \mathbb{R}$ defined by $f_n(t) = nt^2(1 + nt^2)^{-1}$ vanishes at the orgin, and is therefore uniformly approximable on $\mathrm{sp}(x)$ by polynomials of the form $a_1t + a_2t^2 + \cdots + a_kt^k$. It follows that e_n belongs to the closed linear span of $x, x^2, x^3, \ldots$, and in particular, $e_n \in J$.

e_n is clearly self-adjoint, and since the range of each f_n is contained in the unit interval, it follows that $\mathrm{sp}(e_n) \subseteq [0, 1]$.

For (ii), notice that the spectrum of $e - e_n$ is also contained in the unit interval, and so $\|e - e_n\| \leqslant 1$ because $e - e_n$ is self-adjoint. Moreover, the nonnegative function

$$t^2(1 - f_n(t)) = t^2(1 + nt^2)^{-1}$$

is bounded by $1/n$, so that

$$\begin{aligned}\|xe_n - x\|^2 &= \|x(e - e_n)\|^2 = \|(e - e_n)x^2(e - e_n)\| \\ &\leqslant \|x^2(e - e_n)\| \leqslant 1/n.\end{aligned}$$

Therefore $\lim\|xe_n - x\| = 0$.

If $x \neq x^*$, then we may apply the above to x^*x, obtaining e_n such that $\|x^*xe_n - x^*x\| \to 0$. It follows that

$$\begin{aligned}\|xe_n - x\|^2 &= \|x(e_n - e)\|^2 \\ &= \|(e_n - e)x^*x(e_n - e)\|^2 \leqslant \|x^*x(e_n - e)\| \to 0\end{aligned}$$

as $n \to \infty$, as required.

The case where A does not contain an identity is easily dealt with by adjoining an identity, and is left for the reader. □

Corollary 1. *Every ideal in a C^*-algebra is self-adjoint (i.e., is closed under the $*$-operation).*

Proof. Let J be an ideal in a C^*-algebra A, and let x be an element of J. By 1.3.1 we can find a sequence $e_n = e_n^*$ in J so that $x = \lim_n xe_n$. By taking adjoints we have $x^* = \lim e_nx^*$, and clearly the right side of that expression belongs to J. □

Now if J is an ideal in a C^*-algebra A, then the quotient A/J becomes a Banach algebra in the usual way; for example, the norm of the coset representative $\dot{x}$ of x is defined as $\|\dot{x}\| = \inf\{\|x + z\|; z \in J\}$. Because of Corollary 1 above, we may introduce a natural involution in A/J by taking $\dot{x}^*$ to be the coset representative of x^*. It is significant that the norm in A/J is a C^*-norm relative to this involution.

Corollary 2. *A/J is a C^*-algebra.*

PROOF. By Exercise 1.1.B, it suffices to show that $\|\dot{x}\|^2 \leqslant \|\dot{x}^*\dot{x}\|$, for every x in A.

For that, fix x, and let E denote the set of all self-adjoint elements u of J satisfying $\operatorname{sp}(u) \subseteq [0, 1]$. We claim first that

$$\|\dot{x}\| = \inf_{u \in E} \|x - xu\|.$$

Indeed, the inequality $\leqslant$ is obvious because $xu \in J$ for each $u \in E$, and it suffices to show that for each $k \in J$, there is a sequence $u_n \in E$ satisfying $\|x + k\| \geqslant \inf_n \|x - xu_n\|$. Fix k, and choose u_n for k as Proposition 1.3.1. We have already seen that $\|e - u_n\| \leqslant 1$, so that

$$\begin{aligned}\|x + k\| &\geqslant \liminf_n \|(x + k)(e - u_n)\| \\ &= \liminf_n \|x(e - u_n) + k(e - u_n)\| \\ &= \liminf_n \|x - xu_n\| \geqslant \inf_n \|x - xu_n\|,\end{aligned}$$

because $k(e - u_n) = k - ku_n$ tends to zero as $n \to \infty$.

To complete the proof, we apply the preceding formula twice to obtain

$$\begin{aligned}\|\dot{x}\|^2 &= \inf_{u \in E} \|x(e - u)\|^2 = \inf_u \|(e - u)x^*x(e - u)\| \\ &\leqslant \inf_u \|x^*x(e - u)\| = \|\dot{x}^*\dot{x}\|.\end{aligned}$$ □

Now let A and B be C^*-algebras and let π be a $*$-homomorphism of A into B, that is, π preserves the algebraic operations and $\pi(x^*) = \pi(x)^*$. Note that we do not assume that π is bounded, but nevertheless that turns out to be true. To see why, consider first the case where *both* A and B have identities and π maps e_A to e_B. Then clearly π must map invertible elements of A to invertible elements of B, and this implies that π must shrink spectra. Moreover, since the norm of a self-adjoint element of a C^*-algebra must equal its spectral radius, we have

$$\|\pi(x^*x)\| = r(\pi(x^*x)) \leqslant r(x^*x) = \|x^*x\|.$$

Since the left side of this inequality is $\|\pi(x)^*\pi(x)\| = \|\pi(x)\|^2$ and the right side is $\|x\|^2$, we conclude that π has norm at most 1.

Now suppose, in addition to the above hypotheses, that π has trivial kernel. Then we claim that π is *isometric*. By reasoning as above, this will follow if we show that, for every self-adjoint element z of A, $\pi(z)$ and z have the same spectrum. We know that $\operatorname{sp} \pi(z) \subseteq \operatorname{sp}(z)$, so if these sets are different then one can find a continuous function $f:\operatorname{sp}(z) \to \mathbb{R}$ such that $f \neq 0$ but $f = 0$ on $\operatorname{sp} \pi(z)$. Now if f is a polynomial then we have $f(\pi(z)) = \pi(f(z))$. In general, f is the norm limit on $\operatorname{sp}(z)$ of a sequence of polynomials (by the Weierstrass theorem), and so by 1.1.1 we conclude that $f(z)$ and $f(\pi(z))$ are the corresponding limits of polynomials. This proves the formula $f(\pi(z)) = \pi(f(z))$ for arbitrary continuous f. But $f(\pi(z))$ is 0 because $f = 0$ on sp $\pi(z)$, so by the formula we have $\pi(f(z)) = 0$. Since π is assumed injective we conclude that $f(z) = 0$, and by 1.1.1 it follows that $f = 0$ on sp(z), a contradiction.

Note finally that the preceding implies that the range of π is closed even when π is *not* injective. For if we let J be the kernel of π, then π lifts in the obvious way to an injective $\star$-homomorphism $\dot{\pi}$ of A/J into the range of π, and of course $\dot{\pi}$ also preserves identities. Thus $\dot{\pi}$ is isometric by the above, and in particular its range is closed.

To summarize, we have proved the following theorem, at least in the presence of certain assumptions about units.

Theorem 1.3.2. *Let A and B be $C^\star$-algebras and let π be a $\star$-homomorphism of A into B. Then π is continuous and $\pi(A)$ is a $C^\star$-subalgebra of B. π induces an isometric $\star$-isomorphism of the quotient $A/\ker \pi$ onto $\pi(A)$.*

PROOF. We shall merely indicate how the general case can be reduced to the above situation where both A and B have units and $\pi(e_A) = e_B$.

Assume first that A has a unit e_A. By passing from B to the closure of $\pi(A)$ if necessary, we may assume that $\pi(A)$ is dense in B. Since $\pi(e_A)$ is a unit for $\pi(A)$, it is therefore a unit for B, and we are now in the case already discussed.

So assume A has no unit, and let $A_1 \supseteq A$ be the $C^\star$-algebra obtained from A by adjoining a unit e. By adjoining a unit to B if necessary, we may also assume that B has a unit e_B. Define a map $\tilde{\pi}: A_1 \to B$ by

$$\tilde{\pi}(\lambda e + x) = \lambda e_B + \pi(x), \qquad x \in A, \lambda \in \mathbb{C}.$$

It is a simple matter to check that $\tilde{\pi}$ is a $\star$-homormorphism, and $\tilde{\pi}$ is clearly an extension of π to A_1. Moreover, $\tilde{\pi}(e) = e_B$, and thus we are again reduced to the preceding situation. □

Corollary. *Let A be a $C^\star$-algebra and let $|x|$ be another Banach algebra norm on A satisfying $|x^\star x| = |x|^2$, $x \in A$. Then $|x| = \|x\|$ for every $x \in A$.*

PROOF. Let $B = A$, regarded as a $C^\star$-algebra in the norm $|x|$. Then the identity map is an injective $\star$-homomorphism of A on B. Now apply 1.3.2. □

The corollary shows that there is at most one way of making a complex algebra with involution into a $C^\star$-algebra.

We come now to the central concept of this book.

Definition 1.3.3. A *representation of a $C^\star$-algebra A* is a $\star$-homomorphism of A into the $C^\star$-algebra $\mathscr{L}(\mathscr{H})$ of all bounded operators on some Hilbert space $\mathscr{H}$.

It is customary to refer to $\pi: A \to \mathscr{L}(\mathscr{H})$ as a representation of A *on* $\mathscr{H}$. π is called *nondegenerate* if the C^*-algebra of operators $\pi(A)$ has trivial null space. We leave it for the reader to show that, since $\pi(A)$ is self-adjoint, this is equivalent to the assertion that the closed linear span $[\pi(A)\mathscr{H}]$ of all vectors of the form $\pi(x)\xi$, $x \in A$, $\xi \in \mathscr{H}$, is all of $\mathscr{H}$.

An invariant subspace $\mathscr{M}$ for the C^*-algebra $\pi(A)$ is called a *cyclic* subspace if it contains a vector ξ such that the vectors of the form $\pi(x)\xi$, $x \in A$, are dense in $\mathscr{M}$: this is written $\mathscr{M} = [\pi(A)\xi]$. π is called a *cyclic representation* if $\mathscr{H}$ itself is a cyclic subspace for π. It is clear from the preceding paragraph that a cyclic representation is nondegenerate. More generally, a representation of A on $\mathscr{H}$ is nondegenerate if, and only if, $\mathscr{H}$ can be decomposed into a mutually orthogonal family of cyclic subspaces (Exercise 1.3.F).

Let π and σ be two representations of A, perhaps acting on different spaces $\mathscr{H}$ and $\mathscr{K}$. π and σ are said to be *equivalent* if there is a unitary operator $U: \mathscr{H} \to \mathscr{K}$ such that $\sigma(x) = U\pi(x)U^*$ for all x in A; this relation is written $\pi \sim \sigma$. Equivalent representations are indistinguishable in the sense that any geometric property of one must also be shared by the other, and it is correct to think of the unitary operator U as representing nothing more than a change of "coordinates."

Finally, a nonzero representation π of A is called *irreducible* if $\pi(A)$ is an irreducible operator algebra, i.e., commutes with no nontrivial (self-adjoint) projections. Because $\pi(A)$ is a C^*-algebra, this is the same as saying $\pi(A)$ has no nontrivial closed invariant subspaces (Exercise 1.3.D).

Now if A is commutative then so is every image of A under a representation, and it is a simple application of the spectral theorem to see that commutative C^*-algebras of operators on Hilbert spaces of dimension greater than 1 cannot be irreducible (Exercise 1.3.E). So the only irreducible representations of A are those of the form $\pi(x) = \omega(x)I$, where I is the identity operator on a one-dimensional space and ω is a nonzero homomorphism of A into the complex numbers. This shows that we can identify the equivalence classes of irreducible representations of a commutative C^*-algebra in a bijective way with its set of nonzero complex homomorphisms. Moreover, it suggests that one should view an irreducible representation (more precisely, an equivalence class of them) of a noncommutative C^*-algebra as filling a role similar to that of complex homomorphisms. This analogy will be pursued to considerable lengths throughout the book. We will find that while the generalization achieves some remarkable successes within the class of GCR algebras (defined in Section 1.5), it also leads to unexpected and profound difficulties in all other cases.

Returning now to the present discussion, we want to consider a useful connection between representations and ideals. In general, a representation of a C^*-subalgebra of A on a Hilbert space $\mathscr{H}$ cannot be extended to a representation of A on $\mathscr{H}$. But if the subalgebra is an ideal then it can. To see why, let J be an ideal in A and let π be a *nondegenerate* representation of J on a Hilbert space $\mathscr{H}$. We claim first that for each x in A, there is a unique bounded linear operator $\tilde{\pi}(x)$ on $\mathscr{H}$ satisfying $\tilde{\pi}(x)\pi(y) = \pi(xy)$ for every $y \in J$. Indeed, uniqueness is clear from the fact that vectors of the form $\pi(y)\xi$, $y \in J$, $\xi \in \mathscr{H}$, span $\mathscr{H}$. For existence, consider first the case where π is cyclic, and let $\xi_0 \in \mathscr{H}$ be such that $[\pi(J)\xi_0] = \mathscr{H}$. We claim that $\|\pi(xy)\xi_0\| \leqslant \|x\|\cdot\|\pi(y)\xi_0\|$, for each $y \in J$. To see this fix y, and choose a sequence $e_n = e_n^* \in J$ with $\mathrm{sp}(e_n) \subseteq [0, 1]$ and $y^*e_n \to y^*$ (by 1.3.1). By taking adjoints we see that $e_n y \to y$, so that

$$\begin{aligned}\|\pi(xy)\xi_0\| &= \lim_n \|\pi(xe_n y)\xi_0\| \\ &= \lim_n \|\pi(xe_n)\pi(y)\xi_0\| \leqslant \sup_n \|xe_n\|\cdot\|\pi(y)\xi_0\| \leqslant \|x\|\cdot\|\pi(y)\xi_0\|,\end{aligned}$$

as asserted. It follows that the map $\pi(y)\xi_0 \to \pi(xy)\xi_0 (y \in J)$ extends uniquely to an operator $\tilde{\pi}(x)$ on $[\pi(I)\xi_0] = \mathscr{H}$ having norm at most $\|x\|$. The reader can easily check that the required relation $\tilde{\pi}(x)\pi(y) = \pi(xy)$ holds on all vectors of the form $\pi(z)\xi_0$, $z \in J$, so it holds throughout $\mathscr{H}$.

In the general (noncyclic) case, one may apply Exercise 1.3.F to express $\mathscr{H}$ as an orthogonal sum of cyclic subspaces, define $\tilde{\pi}(x)$ as above on each cyclic summand and then add up the pieces in the obvious way to obtain an operator on all of $\mathscr{H}$.

Thus we have established that there is a unique mapping $\tilde{\pi}$ of A into $\mathscr{L}(\mathscr{H})$ which satisfies $\tilde{\pi}(x)\pi(y) = \pi(xy)$ for $x \in A$, $y \in J$. This formula itself implies that $\tilde{\pi}$ preserves the algebraic operations, the involution, and restricts to π on J (these routine verifications are left for the reader). Finally, if σ is any other representation of A on $\mathscr{H}$ such that $\sigma|_J = \pi$, then for every $x \in A$, $y \in J$ we have

$$\sigma(x)\pi(y) = \sigma(x)\sigma(y) = \sigma(xy) = \pi(xy),$$

so that $\sigma = \tilde{\pi}$ by the uniqueness assertion of the preceding paragraph.

If π is a degenerate representation of J on $\mathscr{H}$, then we can make π nondegenerate by passing from $\mathscr{H}$ to the subspace $[\pi(J)\mathscr{H}]$. So we can still obtain a unique extension $\tilde{\pi}$ of π to A such that $\tilde{\pi}(A)$ acts on $[\pi(J)\mathscr{H}]$.

Now suppose, on the other hand, that we start with a representation π of the full algebra A on $\mathscr{H}$. Choose any ideal J in A and let $\mathscr{H}_J = [\pi(J)\mathscr{H}]$. Since J is an ideal we have $\pi(A)\mathscr{H}_J \subseteq \mathscr{H}_J$, and thus $\mathscr{H} = \mathscr{H}_J \oplus \mathscr{H}_J^\perp$ gives a decomposition of $\mathscr{H}$ into reducing subspaces for $\pi(A)$. Define representations π_J and σ_J of A on $\mathscr{H}_J$ and $\mathscr{H}_J^\perp$ respectively by $\pi_J(x) = \pi(x)|_{\mathscr{H}_J}$ and $\sigma_J(x) = \pi(x)|_{\mathscr{H}_J^\perp}$. Then in an obvious sense we have a decomposition $\pi(x) = \pi_J(x) \oplus \sigma_J(x)$ of π, where on the one hand π_J is determined uniquely by the action of π on the ideal J in the sense of the preceding paragraph, and where σ_J annihilates J (recall that $[\pi(J)\mathscr{H}]$ is the null space of the C^*-algebra $\pi(J)$) and can therefore be regarded as a representation of the quotient A/J. These remarks show that once we know all of the representations of both an ideal J and its quotient A/J, then we can reconstruct the representations of A. This procedure is particularly useful when dealing with irreducible representations.

Theorem 1.3.4. *Let π be an irreducible representation of A and let J be an ideal in A such that $\pi(J) \neq 0$. Then $\pi|_J$ is an irreducible representation of J. Every irreducible representation of J extends uniquely to an irreducible representation of A, and if two such representations of J are equivalent then so are their extensions.*

PROOF. Let $\pi: A \to \mathscr{L}(\mathscr{H})$ be irreducible, such that $\pi(J) \neq 0$. We claim that $\pi(J)$ is irreducible. For that it suffices to show that $[\pi(J)\xi] = \mathscr{H}$ for every $\xi \neq 0$ in $\mathscr{H}$. Since $[\pi(J)\xi]$ is invariant under the irreducible C^*-algebra $\pi(A)$, $[\pi(J)\xi]$ must be $\mathscr{H}$ or $\{0\}$. If it is $\{0\}$ then ξ belongs to the null space of the C^*-algebra $\pi(J)$ and therefore $\xi \perp [\pi(J)\mathscr{H}]$. Thus $[\pi(J)\mathscr{H}] \neq \mathscr{H}$ is $\pi(A)$-invariant, and therefore $[\pi(J)\mathscr{H}] = 0$. This means $\pi(J) = 0$, a contradiction.

If, conversely, π is an irreducible representation of J, then its extension $\tilde{\pi}$ must be irreducible because $\tilde{\pi}(A)$ contains the irreducible subalgebra $\pi(J)$.

Finally, if π and σ are irreducible representations of J and U is a unitary operator between their respective spaces such that $\sigma = U\pi U^*$, then note by uniqueness the extensions $\tilde{\pi}$ and $\tilde{\sigma}$ must also satisfy $\tilde{\sigma} = U\tilde{\pi}U^*$. That completes the proof. □

EXERCISES

1.3.A. Let A be a C^*-algebra, let J be an ideal in A, and let B be a sub C^*-algebra of A. Show that $B + J$ is a sub C^*-algebra of A and that $B + J/J$ and $B/B \cap J$ are canonically $*$-isomorphic.

1.3.B. Let J be an ideal in a C^*-algebra A and let x be a self-adjoint element of A. Show that there is an element $k \in J$ satisfying

$$\|x + k\| = \inf_{\ell \in J} \|x + \ell\|.$$

[*Hint*: use the functional calculus to "truncate" x.]

1.3.C. (Hahn decomposition.) Let x be a self-adjoint element of a C^*-algebra A. Show that there exist self-adjoint elements y_1, y_2 in A satisfying $\operatorname{sp}(y_i) \geqslant 0$,

$$\|y_i\| \leqslant \|x\|, \qquad y_1 y_2 = 0, \qquad \text{and} \qquad x = y_1 - y_2.$$

1.3.D. Let π be a nonzero representation of a C^*-algebra A. Show that π is irreducible iff the operator algebra $\pi(A)$ has no nontrivial closed invariant subspaces.

1.3.E. Show that if π is an irreducible representation of a commutative C^*-algebra on a nonzero Hilbert space $\mathscr{H}$, then $\mathscr{H}$ is one-dimensional. [*Hint*: use the spectral theorem.]

1.3.F. Let π be a representation of a C^*-algebra A on $\mathscr{H}$. Show that π is nondegenerate iff $\mathscr{H}$ can be decomposed into an orthogonal sum of cyclic invariant subspaces.

1.4. C^*-algebras of Compact Operators

One usually regards commutative C^*-algebras (equivalently, C^*-algebras of normal operators) as being the easiest to deal with. However, for many purposes, C^*-algebras of compact operators are more tractable even though they may present a high degree of noncommutativity. We will see in this section, for example, that their representation theory can be completely worked out by more or less elementary methods. That result foreshadows the general theory developed in Chapter 4 for GCR algebras. At the end of this section we prove a Wedderburn-type theorem which makes the structure of such algebras quite transparent.

Let $\mathscr{H}$ be a Hilbert space. An operator $T \in \mathscr{L}(\mathscr{H})$ is *compact* if the image of the unit ball of $\mathscr{H}$ under T has compact closure in the norm topology of $\mathscr{H}$. The set $\mathscr{C}(\mathscr{H})$ of all compact operators on $\mathscr{H}$ is a (closed two-sided) ideal in $\mathscr{L}(\mathscr{H})$ (see Exercise 1.4.A) and is therefore a C^*-algebra in its own right. Let us first recall one or two facts about compact operators. For every compact operator T, the spectrum of T is countable and has no nonzero accumulation point ([27], p. 233). Thus if T is self-adjoint and $\mathrm{sp}(T) = \{\lambda_1, \lambda_2, \ldots\}$, $\lambda_i \in \mathbb{R}$, then $\lambda_i \to 0$ as $i \to \infty$ and the spectral theorem for T takes the form $T = \sum_{n=1}^{\infty} \lambda_n E_n$ where the $\{E_n\}$ are mutually orthogonal spectral projections for T. If $\lambda_n \neq 0$ then the characteristic function of $\{\lambda_n\}$ can be uniformly approximated on $\mathrm{sp}(T)$ by polynomials of the form $p(x) = a_1 x + a_2 x^2 + \cdots + a_k x^k$, and therefore E_n can be approximated in the operator norm by polynomials $a_1 T + a_2 T^2 + \cdots + a_k T^k$. In particular each projection E_n is compact, and therefore has finite rank (for $\lambda_n \neq 0$). Since $\|\sum_{k=n}^{\infty} \lambda_k E_k\| = \sup_{k \geq n} |\lambda_k| \to 0$ as $n \to \infty$, it follows that the finite sums $\sum_{k=1}^{n} \lambda_k E_k$ converge in norm to T, and in particular T is a norm limit of self-adjoint operators having finite rank.

Now let $\mathscr{A}$ be a C^*-subalgebra of $\mathscr{C}(\mathscr{H})$, fixed throughout the remainder of this section. By cutting down to an $\mathscr{A}$-invariant subspace, if necessary, we may assume that $\mathscr{A}$ has trivial null space, and thus $[\mathscr{A}\mathscr{H}] = \mathscr{H}$. Note that while $\mathscr{A}$ does not contain the identity if $\mathscr{H}$ is infinite-dimensional, the preceding paragraph shows that it does contain many finite-rank projections, which can be obtained as spectral projections of self-adjoint operators in $\mathscr{A}$. A projection E in $\mathscr{A}$ is called *minimal* if $E \neq 0$ and the only subprojections of E in $\mathscr{A}$ are 0 and E.

Lemma 1.4.1. *Let E be a nonzero projection in $\mathscr{A}$. Then E is minimal iff $E\mathscr{A}E = \{\lambda E : \lambda \in \mathbb{C}\}$. Every nonzero projection in $\mathscr{A}$ is finite-dimensional, and is a finite sum of orthogonal minimal projections.*

PROOF. If $E\mathscr{A}E$ consists of scalar multiples of E, then E is minimal. Conversely, assume E is minimal. It suffices to show that ETE is a scalar multiple of E, for every self-adjoint $T \in \mathscr{A}$. Considering the spectral formula for

ETE we have $ETE = \sum_n \lambda_n F_n$ where the F_n are mutually orthogonal spectral projections of ETE. Since ETE annihilates $E^\perp\mathscr{H}$ so does each F_n, and hence $F_n \leqslant E$. Thus each nonzero F_n must be E, and hence $ETE = \lambda E$ has the required form.

It is plain that every projection in $\mathscr{A}$ is finite-dimensional, by compactness, and the last phrase follows from the usual sort of finite induction. □

Theorem 1.4.2. *If $\mathscr{A}$ is irreducible then $\mathscr{A} = \mathscr{C}(\mathscr{H})$.*

PROOF. Note first that $\mathscr{A}$ contains a projection of rank 1. For if F is a nonzero spectral projection of any self-adjoint operator in $\mathscr{A}$ then by the lemma F contains a minimal projection E, and it suffices to show that E has rank 1. Choose $\xi, \eta \in E\mathscr{H}$, with $\xi \neq 0$ and $\eta \perp \xi$. If T is any operator in $\mathscr{A}$ then ETE has the form λE, by 1.4.1, and thus $(\eta, T\xi) = (\eta, ETE\xi) = (\eta, \lambda\xi) = \bar{\lambda}(\eta, \xi) = 0$, thus $\eta \perp [\mathscr{A}\xi] = \mathscr{H}$ and so $\eta = 0$, which proves $E\mathscr{H} = [\xi]$.

Next observe that $\mathscr{A}$ contains *every* projection F of rank 1. Indeed, such an F has the form $F\xi = (\xi, f)f$ where f is a unit vector in $\mathscr{H}$; so if E is any rank 1 projection in $\mathscr{A}$, say $E\xi = (\xi, e)e$, then we can find a sequence $T_n \in \mathscr{A}$ such that $T_n e \to f$ and $\|T_n e\| = 1$ for every n. Thus $T_n E T_n^* \in \mathscr{A}$, and we have

$$\|T_n E T_n^* \xi - F\xi\| = \|(\xi, T_n e)T_n e - (\xi, f)f\| \leqslant 2\|\xi\| \cdot \|T_n e - f\|,$$

for every $\xi \in \mathscr{H}$. It follows that $\|T_n E T_n^* - F\| \to 0$, proving the assertion. We may now conclude that $\mathscr{A}$ contains every finite rank projection, therefore every self-adjoint compact operator (spectral theorem), and thus $\mathscr{A} = \mathscr{C}(\mathscr{H})$. □

Corollary 1. *$\mathscr{C}(\mathscr{H})$ contains no ideals other than 0 and $\mathscr{C}(\mathscr{H})$.*

PROOF. Let $\mathscr{A} \neq 0$ be an ideal in $\mathscr{C}(\mathscr{H})$. Applying 1.3.4 to the identity representation of $\mathscr{C}(\mathscr{H})$ we see that $\mathscr{A}$ is irreducible, and 1.4.2 shows that $\mathscr{A} = \mathscr{C}(\mathscr{H})$. □

Corollary 2. *Let $\mathscr{B}$ be an irreducible C^*-algebra of operators on $\mathscr{H}$ which contains a nonzero compact operator. The $\mathscr{B}$ contains $\mathscr{C}(\mathscr{H})$.*

PROOF. $\mathscr{B} \cap \mathscr{C}(\mathscr{H})$ is a nonzero ideal in $\mathscr{B}$. Arguing as in Corollary 1 we see that $\mathscr{B} \cap \mathscr{C}(\mathscr{H})$ is irreducible, and by 1.4.2 $\mathscr{B} \cap \mathscr{C}(\mathscr{H}) = \mathscr{C}(\mathscr{H})$, as asserted. □

Proposition 1.4.3. *Let E be a minimal projection in $\mathscr{A}$, let ξ be a unit vector in $E\mathscr{H}$, and let $\mathscr{H}_0 = [\mathscr{A}\xi]$. Then $\mathscr{A}|_{\mathscr{H}_0}$ is irreducible, and in fact $\mathscr{A}|_{\mathscr{H}_0} = \mathscr{C}(\mathscr{H}_0)$.*

PROOF. The map $T \to T|_{\mathscr{H}_0}$ is a $*$-homomorphism of $\mathscr{A}$ into $\mathscr{C}(\mathscr{H}_0)$ whose range is $\mathscr{A}|_{\mathscr{H}_0}$. By 1.3.2 $\mathscr{A}|_{\mathscr{H}_0}$ is a C^*-subalgebra of $\mathscr{C}(\mathscr{H}_0)$, and by 1.4.2 the entire conclusion will follow provided we show that $\mathscr{A}|_{\mathscr{H}_0}$ is irreducible.

For that, choose any $R \in \mathscr{L}(\mathscr{H}_0)$ which commutes with $\mathscr{A}|_{\mathscr{H}_0}$. We will show that R is a scalar. By replacing R with $R - (R\xi, \xi)I$ we can assume

$(R\xi, \xi) = 0$, and in this case we claim $R = 0$. Now if $S, T \in \mathscr{A}$ then by 1.4.1. ET^*SE has the form λE for some $\lambda \in \mathbb{C}$. Thus $(RS\xi, T\xi) = (RSE\xi, TE\xi) = (ET^*RSE\xi, \xi) = (RET^*SE\xi, \xi) = \lambda(RE\xi, \xi) = \lambda(R\xi, \xi) = 0$. $R = 0$ follows because $\mathscr{A}\xi$ is dense in $\mathscr{H}_0$. □

Let π be a nondegenerate representation of $\mathscr{A}$ on a Hilbert space $\mathscr{K}$, and let $\mathscr{K}_0$ be a subspace of $\mathscr{K}$ invariant under $\pi(\mathscr{A})$. Then $\pi_0(T) = \pi(T)|\mathscr{K}_0$ defines a nondegenerate representation of $\mathscr{A}$ on $\mathscr{K}_0$. Such a π_0 is called a *subrepresentation* of π (this is written $\pi_0 \leqslant \pi$). If $\{\pi_i\}$ is any family of subrepresentations of π whose respective subspaces are mutually orthogonal, then $\sum \pi_i$ has an obvious meaning (its value at $T \in \mathscr{A}$ is $\sum \pi_i(T)$), and defines another subrepresentation of π. Now 1.4.3, along with a simple argument, implies that the identity representation of $\mathscr{A}$ is the sum of mutually orthogonal irreducible subrepresentations. The following result is somewhat stronger.

Theorem 1.4.4. *Let π be any nondegenerate representation of $\mathscr{A}$. Then there is an orthogonal family $\{\pi_i\}$ of irreducible subrepresentations of π such that $\pi = \sum_i \pi_i$, and each π_i is equivalent to a subrepresentation of the identity representation of $\mathscr{A}$.*

PROOF. First note that there is a minimal projection $E \in \mathscr{A}$ such that $\pi(E) \neq 0$. To see that, choose $T = T^* \in \mathscr{A}$ such that $\pi(T) \neq 0$. By our initial remarks about the spectral theorem there is a spectral projection F of T such that $\pi(F) \neq 0$. The assertion now follows from the second part of 1.4.1.

Choose such an E. By 1.4.1 there is a linear functional f on $\mathscr{A}$ such that $ETE = f(T)E$, for all $T \in \mathscr{A}$. Let $\mathscr{K}$ be the space on which π acts, and choose a unit vector η(resp. ξ) in $\pi(E)\mathscr{K}$(resp. $E\mathscr{H}$). Then $\mathscr{K}_0 = [\pi(\mathscr{A})\eta]$ defines a subrepresentation π_0 of π. We will show that π_0 is equivalent to the irreducible subrepresentation of the identity representation of $\mathscr{A}$ defined by the subspace $[\mathscr{A}\xi]$ (1.4.3). Indeed, for $T \in \mathscr{A}$ we have

$$\begin{aligned} \|\pi(T)\eta\|^2 &= \|\pi(T)\pi(E)\eta\|^2 = \|\pi(TE)\eta\|^2 \\ &= (\pi(ET^*TE)\eta, \eta) = f(T^*T)(\pi(E)\eta, \eta) \\ &= f(T^*T) = (ET^*TE\xi, \xi) = \|T\xi\|^2. \end{aligned}$$

This shows that map $U: T\xi \mapsto \pi(T)\eta$ extends uniquely to a unitary map of $[\mathscr{A}\xi]$ onto $[\pi(\mathscr{A})\eta]$, and the formula $UT = \pi_0(T)U$ is immediate from the definition of π. That proves the assertion about π_0.

Thus we have proved that every nondegenerate representation of $\mathscr{A}$ contains an irreducible nonzero subrepresentation equivalent to a subrepresentation of id. The proof of the theorem can now be completed by an exhaustion argument (i.e., by Zorn, choose a maximal family $\{\pi_i\}$ of orthogonal subrepresentations of π each of which is equivalent to an irreducible subrepresentation of id, and note that $\sum \pi_i$ must be π by maximality). □

Let π be a representation of $\mathscr{A}$ on $\mathscr{K}$, and let n be a positive cardinal. Let $\mathscr{K}'$ be the direct sum of n copies of $\mathscr{K}$. Then we can define a representation

$n \cdot \pi$ of $\mathscr{A}$ on $\mathscr{K}'$ by setting $(n \cdot \pi)(T) = \sum^{\oplus} \pi(T)$, the direct sum extended over n copies of $\pi(T)$. $n \cdot \pi$ is called a *multiple* of π.

Corollary 1. *Every representation of $\mathscr{C}(\mathscr{H})$ is equivalent to a multiple of the identity representation.*

PROOF. Let π be a representation of $\mathscr{A}$ on $\mathscr{K}$. Since the identity representation *id* of $\mathscr{C}(\mathscr{H})$ is itself irreducible, it follows that π has a decomposition $\sum_i \pi_i$ into orthogonal subrepresentations each of which is equivalent to *id*. Choosing, for each i, a unitary operator U_i from $\mathscr{H}$ to the range space of $\mathscr{H}_i$ such that $id = U_i^* \pi_i U_i$ and letting n be the cardinal of $\{\pi_i\}$, we see that $U = \sum_i^{\oplus} U_i$ implements the equivalence $n \cdot id = U^* \pi U$. □

Corollary 2. *Every irreducible representation of $\mathscr{C}(\mathscr{H})$ is equivalent to the identity representation.*

Corollary 3. *Let $\mathscr{H}$ and $\mathscr{K}$ be Hilbert spaces. Then every $*$-isomorphism α of $\mathscr{C}(\mathscr{H})$ (resp. $\mathscr{L}(\mathscr{H})$) onto $\mathscr{C}(\mathscr{K})$ (resp. $\mathscr{L}(\mathscr{K})$) is implemented by a unitary operator U; $\alpha(T) = UTU^*$.*

PROOF. Consider first the case $\alpha: \mathscr{C}(\mathscr{H}) \to \mathscr{C}(\mathscr{K})$. Then α is an irreducible representation of $\mathscr{C}(\mathscr{H})$, and the conclusion follows from Corollary 2.

Now suppose α is a $*$-isomorphism of $\mathscr{L}(\mathscr{H})$ on $\mathscr{L}(\mathscr{K})$ and let $\alpha_0 : \mathscr{C}(\mathscr{H}) \to \mathscr{L}(\mathscr{K})$ be the restriction of α to $\mathscr{C}(\mathscr{H})$. Since α is an irreducible representation and $\mathscr{C}(\mathscr{H})$ is an ideal in $\mathscr{L}(\mathscr{H})$, it follows by 1.3.4 that α_0 is irreducible. By Corollary 2 α_0 has the form $\alpha_0(T) = UTU^*$, where U is a unitary operator from $\mathscr{H}$ to $\mathscr{K}$. Since $\beta(T) = UTU^*$, $T \in \mathscr{L}(\mathscr{H})$, is an irreducible representation of $\mathscr{L}(\mathscr{H})$ extending α_0, we conclude from 1.3.4. that $\beta = \alpha$, as required. □

We now wish to restate 1.4.4 in a more systematic way. The set of all irreducible subrepresentations of the identity representation of $\mathscr{A}$ is partitioned by the unitary equivalence relation $\sim$. Let $\hat{\mathscr{A}}$ denote the set of all these equivalence classes. $\hat{\mathscr{A}}$ is called the *spectrum* of $\mathscr{A}$, and here $\hat{\mathscr{A}}$ is to be regarded as a set with no additional structure. For each $\zeta \in \hat{\mathscr{A}}$ choose, once and for all, an element $\pi_\zeta \in \zeta$. It will be convenient to write $\zeta(T)$ for the operator $\pi_\zeta(T)$, $T \in \mathscr{A}$. Now if $\{\pi_i\}$ is a family of irreducible representations of $\mathscr{A}$ such that each π_i is equivalent to ζ, then as we have seen in the proof of Corollary 1, the direct sum $\sum_i^{\oplus} \pi_i$ is equivalent to $n \cdot \zeta$ where n is the cardinal number of $\{\pi_i\}$. Thus 1.4.4 can be paraphrased as follows: *every nondegenerate representation of $\mathscr{A}$ is equivalent to a representation of the form $\sum^{\oplus} n(\zeta) \cdot \zeta$, the sum extended over all classes $\zeta \in \hat{\mathscr{A}}$, where $\zeta \mapsto n(\zeta)$ is a function from $\hat{\mathscr{A}}$ into the class of all nonnegative cardinal numbers.* Now it can be shown that if $\zeta \mapsto m(\zeta)$ is another cardinal-valued function, then $\sum^{\oplus} m(\zeta) \cdot \zeta$ and $\sum^{\oplus} n(\zeta) \cdot \zeta$ are equivalent if, and only if, $m(\zeta) = n(\zeta)$ for all $\zeta \in \hat{\mathscr{A}}$ (while one can give an elementary proof in this case because $\mathscr{A} \subseteq \mathscr{C}(\mathscr{H})$ we shall not do so, because a more general theorem will be proved later, in Chapters 2 and 4). In particular, the function $n(\cdot)$ occurring in the expression $\pi = \sum^{\oplus} n(\zeta) \cdot \zeta$ is well-defined by π; $n(\zeta)$ is called the *multiplicity* of ζ in π, and n itself is called the *multiplicity function* of π. Thus, two representations of $\mathscr{A}$ are equivalent iff they have the same multiplicity functions, and we have here an effective classification of the representations of $\mathscr{A}$.

Let us apply this to single operators. Choose a compact operator $T \in \mathscr{C}(\mathscr{H})$, and let $\mathscr{A}$ be the C^*-algebra generated by T. For every $\zeta \in \hat{\mathscr{A}}$ define T_ζ as $\zeta(T)$. T_ζ is clearly compact and, since it generates the irreducible C^*-algebra $\zeta(\mathscr{A})$, it is also irreducible. Moreover, T_ζ and $T_{\zeta'}$ are not unitarily equivalent for $\zeta \neq \zeta'$, because any unitary operator U satisfying $UT_\zeta U^* = T_{\zeta'}$ would implement an equivalence between the inequivalent representation ζ and ζ' of $\mathscr{A}$. Applying the above decomposition to the identity representation of $\mathscr{A}$ and evaluating everything at T, we obtain the decomposition $T = \sum^{\oplus} n(\zeta) \cdot T_\zeta$, where $m \cdot T_\zeta$ denotes the direct sum of m copies of T_ζ. Now since each operator $n(\zeta) \cdot T_\zeta$ is compact and nonzero, $n(\cdot)$ must take on *finite* positive values $1, 2, \ldots$. Moreover, the compactness of $\sum^{\oplus} n(\zeta) \cdot T_\zeta$ implies that $\lim_{\zeta \to \infty} \|T_\zeta\| = 0$ in the sense that for every $\varepsilon > 0$ there is a finite subset $E \subseteq \hat{\mathscr{A}}$ such that $\|T_\zeta\| \leqslant \varepsilon$ for $\zeta \notin E$. Conversely, given any family $\{T_i : i \in I\}$ of mutually inequivalent irreducible compact operators such that $\lim_{i \to \infty} \|T_i\| = 0$, and given any set $\{n_i : i \in I\}$ of positive integers, $T' = \sum_i^{\oplus} n_i \cdot T_i$ defines a compact operator. As it turns out, the uniqueness assertion of the preceding paragraph takes the following form: *T' is unitarily equivalent to $T = \sum^{\oplus} n(\zeta) \cdot T_\zeta$ iff there is a bijection $i \in I \mapsto \zeta_i \in \hat{\mathscr{A}}$ such that $n(\zeta_i) = n_i$ and T_{ζ_i} is unitarily equivalent to T_i, for all $i \in I$.*

This gives the precise sense in which the theory of C^*-algebras reduces the problem of classifying general compact operators to the problem of classifying irreducible compact operators. But that is as far as the self-adjoint theory takes us; the classification problem for irreducible compact operators is a separate problem requiring its own techniques [1, 2, 5].

Let us now examine the algebraic structure of a general C^*-algebra of compact operators. An abstract C^*-algebra A is called *elementary* if A is $*$-isomorphic to the C^*-algebra $\mathscr{C}(\mathscr{H})$ of all compact operators on some Hilbert space $\mathscr{H}$. Now for any family $\{A_n : n \in I\}$ of C^*-algebras (the index set I is perhaps uncountable), define the *direct sum* $\sum A_n$ as the set of all functions $n \in I \to a_n \in A_n$ which satisfy the condition $\lim_{n \to \infty} \|a_n\| = 0$ in the sense described above. We can make $\sum A_n$ into a C^*-algebra by giving it the "pointwise" operations (for example, $\{a_n\} + \{b_n\} = \{a_n + b_n\}$) and the norm $\|\{a_n\}\| = \sup_n \|a_n\|$. The *direct product* $\prod A_n$ has a similar definition except that one takes for its elements *all* functions $\{a_n\}$ satisfying $\|\{a_n\}\| = \sup_n \|a_n\| < \infty$. Thus $\prod A_n$ is much larger than $\sum A_n$, except in the case when the index set is finite.

Theorem 1.4.5. *Every C^*-algebra of compact operators is $*$-isomorphic to a direct sum of elementary C^*-algebras.*

PROOF. Let $\mathscr{A}$ be a C^*-subalgebra of $\mathscr{C}(\mathscr{H})$. As before, choose a representative ζ for each equivalence class of irreducible subrepresentations of the identity representation *id* of $\mathscr{A}$. For each ζ, let $\mathscr{H}_\zeta$ be the subspace of $\mathscr{H}$ on which ζ acts. We will show that $\mathscr{A}$ is $*$-isomorphic with $\sum \mathscr{C}(\mathscr{H}_\zeta)$, the sum extended over all $\zeta \in \hat{\mathscr{A}}$.

Now by the preceding discussion we can write id $= \sum^\oplus n(\zeta) \cdot \zeta$, where $n(\zeta)$ is a positive cardinal for every $\zeta \in \hat{\mathscr{A}}$. Thus $T = \sum^\oplus n(\zeta) \cdot \zeta(T)$ for each $T \in \mathscr{A}$. Clearly $\|T\| = \sup_\zeta \|n(\zeta) \cdot \zeta(T)\| = \sup_\zeta \|\zeta(T)\|$, and since T is compact we must have $\lim_{\zeta \to \infty} \|\zeta(T)\| = 0$. Thus the map which associates to each $T \in \mathscr{A}$ the element $\{T_\zeta\}$ of $\sum \mathscr{C}(\mathscr{H}_\zeta)$ defined by $T_\zeta = \zeta(T)$ is an isometric $*$-homomorphic imbedding of $\mathscr{A}$ in $\sum \mathscr{C}(\mathscr{H}_\zeta)$. We have to show that it maps onto $\sum \mathscr{C}(\mathscr{H}_\zeta)$.

Choose an element $\{T_\zeta\} \in \sum \mathscr{C}(\mathscr{H}_\zeta)$. For each ζ, $\zeta(\mathscr{A})$ is an irreducible C^*-subalgebra of $\mathscr{C}(\mathscr{H}_\zeta)$ and so $\zeta(\mathscr{A}) = \mathscr{C}(\mathscr{H}_\zeta)$ by 1.4.2. Thus there is an operator $S_\zeta \in \mathscr{A}$ such that $\zeta(S_\zeta) = T_\zeta$. By 1.3.2 we can choose S_ζ so that $\|S_\zeta\|$ is arbitrarily close to $\|T_\zeta\|$, and in particular we can arrange that $\lim_{\zeta \to \infty} \|S_\zeta\| = 0$. Let E_ζ be the projection of $\mathscr{H}$ onto $[\mathscr{A}'\mathscr{H}_\zeta]$. Since $[\mathscr{A}'\mathscr{H}_\zeta]$ is invariant under both $\mathscr{A}$ and $\mathscr{A}'$ it follows that E_ζ belongs to the center of $\mathscr{A}''$. We claim first, that $S_\zeta E_\zeta \in \mathscr{A}$, and second, that $E_\zeta \perp E_{\zeta'}$ if $\zeta \neq \zeta'$. Granting that, it follows that $S = \sum S_\zeta E_\zeta$ is a well-defined element of $\mathscr{A}$ (because $\|S_\zeta\|$ tends to 0), and we have $\zeta(S) = \zeta(S_\zeta E_\zeta) = \zeta(S_\zeta) = T_\zeta$ (because $E_{\zeta'}$ is orthogonal to $\mathscr{H}_\zeta$, the range of the subrepresentation ζ, for $\zeta' \neq \zeta$). Thus the theorem will be proved.

The first claim follows if we show that $\mathscr{A}T \subseteq \mathscr{A}$ for every self-adjoint $T \in \mathscr{A}''$; or equivalently, that $T\mathscr{A} \subseteq \mathscr{A}$ (since $\mathscr{A}$ is self-adjoint). For such a T, Kaplansky's density theorem provides a bounded net $T_n \in \mathscr{A}$ such that $T_n \to T$ strongly. Thus $\|T_n K - TK\| \to 0$ for every finite rank operator K, and since $T_n - T$ is bounded this persists for arbitrary compact K. Because $\mathscr{A}$ is norm-closed we conclude that $T\mathscr{A} \subseteq \mathscr{A}$.

It remains to prove that $E_\zeta \perp E_{\zeta'}$ if $\zeta \neq \zeta'$. By definition of E_ζ this is equivalent to $\mathscr{A}'\mathscr{H}_\zeta \perp \mathscr{H}_{\zeta'}$ if $\zeta' \neq \zeta$. Assume, to the contrary, that there is an element $T' \in \mathscr{A}'$ such that $F_{\zeta'} T' F_\zeta \neq 0$, F_ζ and $F_{\zeta'}$ denoting the projections on $\mathscr{H}_\zeta$ and $\mathscr{H}_{\zeta'}$. From the polar decomposition of the operator $F_{\zeta'} T F_\zeta \in \mathscr{A}'$ we obtain a nonzero partial isometry $U \in \mathscr{A}'$ whose initial space (resp. final space) is contained in $\mathscr{H}_\zeta$ (resp. $\mathscr{H}_{\zeta'}$) Since these spaces are invariant under the respective irreducible C^*-algebras $\zeta(\mathscr{A})$ and $\zeta'(\mathscr{A})$ we see that $V = U|\mathscr{H}_\zeta$ is a unitary map of $\mathscr{H}_\zeta$ on $\mathscr{H}_{\zeta'}$. Since $V\zeta(T) = \zeta'(T)V$ (because $U \in \mathscr{A}'$) we have a contradiction of the fact that ζ and ζ' are inequivalent. □

1.5. CCR and GCR Algebras

In the preceding section we saw how one can obtain rather complete information about noncommutative C^*-algebras of compact operators. We are now going to introduce a much broader class of C^*-algebras. The ideas, and most of the results of this section, originated in a paper of Kaplansky [15].

Definition 1.5.1. *A* CCR *algebra is a C^*-algebra A such that, for every irreducible representation π of A, $\pi(A)$ consists of compact operators.*

The acronym CCR is supposed to suggest the tortured phrase "completely continuous representations." 1.4.4 implies that every C^*-algebra of compact operators is CCR. At the other extreme, every commutative C^*-algebra is CCR since each of its irreducible representations is one-dimensional. More generally, A is CCR if each of its irreducible representations is finite-dimensional (it is not necessary that they have the same dimension, or that the dimensions are even bounded).

We shall only need one or two results about CCR algebras. Note first that for every irreducible representation π of a CCR algebra A on $\mathscr{H}$, we see by 1.4.2 that $\pi(A)$ must in fact coincide with the algebra $\mathscr{C}(\mathscr{H})$ of all compact operators on $\mathscr{H}$. Since π induces a $*$-isomorphism of $A/\ker \pi$ on $\mathscr{C}(\mathscr{H})$ and since $\mathscr{C}(\mathscr{H})$ has no nontrivial closed ideals, we conclude that 0 is a maximal ideal in $A/\ker \pi$. Equivalently, *the kernel of an irreducible representation of* a CCR *algebra A is a maximal ideal in A.* The reader is cautioned that this conclusion is false for more general C^*-algebras.

Proposition 1.5.2. *Let A be a* CCR *algebra and let π and σ be irreducible representations of A such that* $\ker \pi \subseteq \ker \sigma$. *Then π and σ are equivalent.*

Proof. Suppose π and σ act on Hilbert spaces $\mathscr{H}$ and $\mathscr{K}$. By the preceding remarks $\pi(A) = \mathscr{C}(\mathscr{H})$, and by the hypothesis $\ker \pi \subseteq \ker \sigma$, the mapping $\lambda : \pi(x) \to \sigma(x)$ $(x \in A)$ defines an irreducible representation of $\mathscr{C}(\mathscr{H})$ on $\mathscr{K}$. By Corollary 2 of 1.4.4, λ is unitarily equivalent to the identity representation, and note that the latter implies $\pi \sim \sigma$. □

Note in particular that an irreducible representation of a CCR algebra is determined (to equivalence) by its kernel. As we will see, this useful fact is true in somewhat greater generality (1.5.4).

There are many common C^*-algebras which are not CCR. Here is a simple example. Let T be an irreducible operator on a Hilbert space $\mathscr{H}$ which is not compact but whose imaginary part is compact. Then $C^*(T)$ is not CCR, for the identity representation of $C^*(T)$ is irreducible but its range contains the noncompact operators T and I. Nevertheless, the structure of $C^*(T)$ is not at all pathological. Note for example that $C^*(T)$ contains $\mathscr{C}(\mathscr{H})$, by Corollary 2 of 1.4.2. Since the quotient map of $C^*(T)$ on $C^*(T)/\mathscr{C}(\mathscr{H})$ annihilates the imaginary part of T, $C^*(T)/\mathscr{C}(\mathscr{H})$ is generated by a single self-adjoint element and the identity, and is therefore commutative. Thus both the ideal $\mathscr{C}(\mathscr{H})$ and its quotient $C^*(T)/\mathscr{C}(\mathscr{H})$ are CCR algebras of the most tractable kind.

Now let A be a general C^*-algebra, and let π be an irreducible representation of A on $\mathscr{H}$. Since $\mathscr{C}(\mathscr{H})$ is an ideal in $\mathscr{L}(\mathscr{H})$ it follows that the set $\mathscr{C}_\pi = \{x \in A : \pi(x) \in \mathscr{C}(\mathscr{H})\}$ is an ideal in A, which contains $\ker \pi$. Of course it is possible that $\mathscr{C}_\pi = \ker \pi$. But in any event the intersection CCR(A) of all these ideals $\mathscr{C}_\pi$ as π runs over all irreducible representations is an ideal in A, consisting of those elements of A which are compact in every irreducible representation. Thus by 1.3.4 CCR(A) is a CCR algebra in its own right, and moreover CCR(A) contains every other CCR ideal in A. Thus CCR(A) is the largest CCR ideal in A. Needless to say, CCR(A) might be 0.

Definition 1.5.3. *A* GCR *algebra is a C^*-algebra A such that* CCR$(A/J) \neq 0$ *for every ideal $J \neq A$.*

Recalling 1.3.2 we see that GCR algebras are characterized by the fact that all of their $*$-homomorphic images contain a nonzero CCR ideal. Now, given any quotient A/J of a C^*-algebra A, then we may compose an irreducible representation of A/J with the quotient map of A on A/J to obtain an irreducible representation of A. It follows that *every quotient of a* CCR *algebra is* CCR. In particular, every CCR algebra is GCR (so that GCR algebras are "generalized CCR algebras").

Proposition 1.5.4. *Let A be a* GCR *algebra. Then for every irreducible representation π of A on $\mathscr{H}$, $\pi(A)$ contains $\mathscr{C}(\mathscr{H})$. Two irreducible representations of A which have the same kernel are equivalent.*

PROOF. Now $\pi(A)$ is $*$-isomorphic with $A/\ker \pi$ and therefore it contains a nonzero CCR ideal. Since the identity representation of the ideal is irreducible (1.3.4) this implies in particular that $\pi(A)$ contains a nonzero compact operator. Therefore the first conclusion follows from Corollary 2 of 1.4.2.

Now let π and σ be two irreducible representations of A on spaces $\mathscr{H}$ and $\mathscr{K}$, respectively, such that $\ker \pi = \ker \sigma$. Thus the map $\lambda: \pi(x) \mapsto \sigma(x)$, $x \in A$, defines an irreducible representation of $\pi(A)$. Since $\pi(A)$ contains $\mathscr{C}(\mathscr{H})$, the restriction $\lambda|_{\mathscr{C}(\mathscr{H})}$ defines a nonzero representation of $\mathscr{C}(\mathscr{H})$ which by 1.3.4 is irreducible. By Corollary 2 of 1.4.4 we conclude that $\lambda|_{\mathscr{C}(\mathscr{H})}$ is equivalent to the identity representation of $\mathscr{C}(\mathscr{H})$. Finally, by 1.3.4. again, it follows that λ itself is equivalent to the identity representation of $\pi(A)$. Note that this implies π is equivalent to σ. □

The definition of GCR algebras we have given is not a convenient one to check, for one first has to find all the ideals J in a given algebra A before he can examine the quotients A/J. The definition of CCR algebras is easier to deal with; one simply finds all irreducible representations π of A and verifies that $\pi(A)$ consists of compact operators. The preceding result implies that GCR algebras have an analogous property, namely that for every irreducible representation π, $\pi(A)$ contains at least one nonzero compact operator (and therefore all of them). Thus it is natural to ask if the latter property implies that A is GCR. The answer is yes, but the proof is hard. The conclusion follows (in the separable case) from a deep theorem of Glimm [13], or by a more direct argument given independently by Dixmier [11]. The proof has been extended to the inseparable case by Sakai [24, 25]. It turns out that the second property asserted in 1.5.4 also characterizes separable GCR algebras [13], [11], but that is also hard. In Theorem 1.5.5 below we will give another structural condition which is equivalent to the GCR property.

Consider first the operator T in the discussion preceding 1.5.3. Then the chain of ideals $\{0, \mathscr{C}(\mathscr{H}), C^*(T)\}$ in $C^*(T)$ has the property that the successive quotients $\mathscr{C}(\mathscr{H})/0$, $C^*(T)/\mathscr{C}(\mathscr{H})$, are CCR algebras. This property can be generalized in the following way. A *composition series* in a C^*-algebra A is a family of ideals $\{J_\alpha; 0 \leqslant \alpha \leqslant \alpha_0\}$ indexed by the ordinals α, $0 \leqslant \alpha \leqslant \alpha_0$, having the following properties:

(i) *for all* $\alpha < \alpha_0$, J_α *is contained properly in* $J_{\alpha+1}$;
(ii) $J_0 = 0$, $J_{\alpha_0} = A$;
(iii) *if* β *is a limit ordinal then* J_β *is the norm closure of* $\bigcup_{\alpha<\beta} J_\alpha$.

Thus in the example we have a composition series of length 3, though in general of course a composition series can be infinite.

Theorem 1.5.5. *Every* GCR *algebra* A *has exactly one composition series* $\{J_\alpha : 0 \leqslant \alpha \leqslant \alpha_0\}$ *with the property that* $J_{\alpha+1}/J_\alpha$ *is the largest* CCR *ideal in* A/J_α *for every* α, $0 \leqslant \alpha < \alpha_0$. *Conversely, if* A *admits a composition series* $\{J_\alpha : 0 \leqslant \alpha \leqslant \alpha_0\}$ *such that each quotient* $J_{\alpha+1}/J_\alpha$ *is* CCR, *then* A *is* GCR.

Proof. Let A be a GCR algebra. We define the series $\{J_\alpha\}$ by transfinite induction. Put $J_0 = 0$. Note that $A = A/J_0$ contains a nonzero CCR ideal. Inductively, let β be an ordinal such that J_α has been defined for all $\alpha < \beta$ and satisfies (i), (ii), (iii) above as well as the property stated in the theorem, whenever they make sense. Now if β has an immediate predecessor β_-, and if $J_{\beta_-} = A$, then the construction ends. If $J_{\beta_-} \neq A$, then define J_β to be the set of all $x \in A$ which are mapped into $\mathrm{CCR}(A/J_{\beta_-})$ under the quotient map of A on A/J_{β_-}. Clearly J_β is an ideal containing J_{β_-} for which $J_\beta/J_{\beta_-} = \mathrm{CCR}(A/J_{\beta_-})$ is the largest CCR ideal in A/J_{β_-}. If, on the other hand, β is a limit ordinal, then define J_β as the norm closure of $\bigcup_{\alpha<\beta} J_\alpha$. This defines a composition series $\{J_\alpha : 0 \leqslant \alpha \leqslant \alpha_0\}$ having the required properties. Note for example that the induction must terminate at some ordinal whose cardinal number does not exceed $2^{\mathrm{card}\, A}$.

For uniqueness, let $\{K_\alpha : 0 \leqslant \alpha \leqslant \beta_0\}$ be another such composition series, different from $\{J_\alpha : 0 \leqslant \alpha \leqslant \alpha_0\}$, and assume for definiteness that $\alpha_0 \leqslant \beta_0$. Now if $J_\alpha = K_\alpha$ for all $\alpha \leqslant \alpha_0$ then we have a contradiction, for $J_{\alpha_0} = K_{\alpha_0} = A$ implies that $\beta_0 = \alpha_0$ and the two series are identical. Thus there is a first ordinal $\gamma \leqslant \alpha_0$ such that $J_\gamma \neq K_\gamma$. $\gamma > 0$ because $J_0 = K_0 = 0$. Note also that by property (iii) in the definition of composition series γ cannot be a limit ordinal. Thus γ has an immediate predecessor γ_-, and of course we have $J_{\gamma_-} = K_{\gamma_-}$. Since both quotients J_γ/J_{γ_-} and K_γ/J_{γ_-} equal the largest CCR ideal in A/J_{γ_-} we conclude that $J_\gamma = K_\gamma \pmod{J_{\gamma_-}}$ and hence $J_\gamma = K_\gamma$, a contradiction.

Finally, let $\{J_\alpha : 0 \leqslant \alpha \leqslant \alpha_0\}$ be a composition series for A such that each quotient $J_{\alpha+1}/J_\alpha$ is CCR. We have to prove that for every ideal $K \neq A$, A/K contains a nonzero CCR ideal. Now since $\bigcup_\alpha J_\alpha = A$ it follows that there is a first $\gamma > 0$ such that the ideal J_γ/K in A/K is nonzero. We will show that J_γ/K is a CCR algebra. Again, we may argue as above to see that γ is not a limit ordinal. Thus γ has an immediate predecessor γ_-, for which $J_{\gamma_-}/K = 0$, i.e., $J_{\gamma_-} \subseteq K$. Thus the mapping $x + J_{\gamma_-} \mapsto x + K$, $x \in J_\gamma$, defines a $*$-homomorphism of the CCR algebra J_γ/J_{γ_-} onto J_γ/K, so by 1.3.2 J_γ/K is $*$-isomorphic to a quotient of the CCR algebra J_γ/J_{γ_-}. Thus J_γ/K is CCR. □

We remark that if A is a *separable* GCR algebra then its composition series $\{J_\alpha : 0 \leqslant \alpha \leqslant \alpha_0\}$ must be countable. Indeed we can choose, for each $\alpha < \alpha_0$, an element $x_\alpha \in J_{\alpha+1}$ such that $\|x_\alpha - z\| \geqslant 1$ for all $z \in J_\alpha$. (this is because $J_{\alpha+1}/J_\alpha$ is not 0). Because $\{J_\alpha\}$ is a well-ordered set of ideals it follows easily that $\|x_\alpha - x_\beta\| \geqslant 1$ if $\alpha \neq \beta$. Because A is separable, $\{x_\alpha : 0 \leqslant \alpha < \alpha_0\}$ must be a countable set, and therefore α_0 is a countable ordinal.

Note finally that the last statement of 1.5.5 shows that each operator T of the type discussed above is a GCR operator in the following sense.

Definition 1.5.6. *A Hilbert space operator T is called a* GCR *operator if $C^*(T)$ is a* GCR *algebra.*

Since finitely generated C^*-algebras are always separable, the composition series associated with a GCR operator is countable. The class of all GCR operators contains the normal operators, the compact operators, and a great variety of others as well. Moreover, we will see in the sequel that it is the GCR operators (and they alone) which lend themselves to the possibility of analysis in terms of "generalized" spectral theory and multiplicity theory.

As a concluding remark, it is not hard to see that a general C^*-algebra A contains a unique ideal K such that K is a GCR algebra in its own right and A/K has *no* nonzero CCR ideals (see Exercise 1.5.B). C^*-algebras with the latter property are called NGCR. We shall have little, and certainly nothing good, to say about NGCR algebras here.

Exercises

1.5.A. Show that if J is an ideal in a GCR algebra A, then both J and A/J are GCR algebras

1.5.B. A C^*-algebra having no nonzero CCR ideals is called an NGCR algebra. Show that every C^*-algebra A contains a unique ideal K such that K is GCR and A/K is NGCR.

1.5.C. Let $\mathscr{A} \subseteq \mathscr{L}(\mathscr{H})$ be an irreducible C^*-algebra of operators. Show that A is an NGCR algebra iff $\mathscr{A}$ contains no nonzero compact operators.

1.5.D. Let S be the unilateral shift (cf. Exercise 1.4.D). Show that $C^*(S)$ is a GCR algebra and describe its canonical composition series.

1.5.E. A *unilateral weighted shift* is an operator defined on an orthonormal base $e_1, e_2, \ldots$ by the condition

$$A: e_n \to w_n e_{n+1},$$

where $\{w_n\}$ is a bounded sequence of nonnegative real numbers.

a. Show that if $w_n > 0$ for every n, then $C^*(A)$ is irreducible.

b. Show that if $w_n \geq \delta > 0$ for every n, then $C^*(A)$ cannot be an NGCR algebra. [*Hint*: use 1.2.A and 1.5.C.]

1.6. States and the GNS Construction

We now want to discuss certain matters relating to the existence of representations of C^*-algebras, and how one goes about constructing them. If we are given a C^*-algebra $\mathscr{A}$ of operators on a Hilbert space then there are always certain representations in evidence, namely the identity representation of $\mathscr{A}$ and its subrepresentations. If $\mathscr{A}$ contains only compact operators, then the results of Section 1.4 provide a very explicit method for constructing *all* possible representations of $\mathscr{A}$ from the irreducible subrepresentations of *id* (to be perfectly accurate, we only obtain a representative from each unitary equivalence class, but of course that is good enough). On the other hand, if $\mathscr{A}$ contains some noncompact operators, then this method does not exhaust the possibilities; it can be shown, for example, that in this case there will always exist irreducible representations of $\mathscr{A}$ which are not equivalent to subrepresentations of the identity representation. Indeed, the identity representation of $\mathscr{A}$ might contain no irreducible subrepresentations whatsoever. In the extreme case, the given C^*-algebra may be defined in some abstract fashion which does not put into evidence even a single nontrivial representation.

The purpose of this section is to show how representations of an abstract C^*-algebra can be constructed from certain linear functionals, and in Section 1.7 we will show that these functionals (and therefore representations) always exist in abundance.

Let A be an abstract C^*-algebra with unit e, which will be fixed throughout the discussion. A linear functional f on A is said to be *positive* if $f(z^*z) \geqslant 0$ for every z in A; if f is normalized so that $f(e) = 1$, then it is called a *state*. As an example, suppose we are given a representation π of A on a Hilbert space $\mathscr{H}$ and a vector ξ in $\mathscr{H}$. Then $f(x) = (\pi(x)\xi, \xi)$ defines a linear functional on A, and it is very easy to see that, in fact, f is positive. If π is nondegenerate, then f is a state if and only if ξ is a unit vector. We will first describe a very useful procedure, due to Gelfand, Naimark, and Segal, whereby one starts with a positive functional f and constructs π and ξ so that the above relation is satisfied. We then characterize those states f which give rise to irreducible representations π.

With every linear functional f on A there is an associated sesquilinear form $[\cdot, \cdot]$ on $A \times A$, defined by $[x, y] = f(y^*x)$. Notice that when f is a positive linear functional its associated form is positive semidefinite, and therefore satisfies the Schwarz inequality. In terms of f, this is simply

$$|f(y^*x)|^2 \leqslant f(x^*x)f(y^*y), \tag{1.6.1}$$

for every x, y in A. 1.6.1 is called the Schwarz inequality for positive linear functionals. We shall also require the following fact, which implies that positive linear functionals are automatically continuous.

Proposition 1.6.2. *Every positive linear functional f has norm $f(e)$.*

Proof. We claim first that if x is an element of A satisfying $x = x^*$ and $\|x\| \leqslant 1$, then there is a self-adjoint element $y \in A$ such that $e - x = y^2$. Indeed, the sub C^*-algebra generated by x and e is abelian, and the image of x under the Gelfand map is a real-valued continuous function taking values in the interval $[-1, +1]$. Thus $e - x$ corresponds to a continuous function taking values in $[0, +2]$, which therefore has a continuous real-valued square root. y may be taken as the inverse image of the latter under the Gelfand map.

Now let $z \in A$, $\|z\| \leqslant 1$. Then by the Schwarz inequality 1.6.1 we have $|f(z)|^2 = |f(e^*z)|^2 \leqslant f(z^*z)f(e^*e) = f(z^*z)f(e)$. So to prove $\|f\| \leqslant f(e)$ it suffices to show that $f(z^*z) \leqslant f(e)$, or equivalently, $f(e - z^*z) \geqslant 0$. But by the preceding paragraph we know that there is an element $y = y^*$ in A such that $e - z^*z = y^2$, from which the assertion is evident. The opposite inequality $\|f\| \geqslant f(e)$ is apparent from the fact that e has unit norm. □

So in particular, every state of A has norm 1. It is significant that, while the proof we have given works only for C^*-algebras, 1.6.2 itself is true for a much broader class of Banach $*$-algebras ([6], p. 22). We come now to the main discussion, the result of which is summarized as follows:

Theorem 1.6.3. *For every positive linear functional f on A, there is a representation π of A and a vector ξ such that*

$$f(x) = (\pi(x)\xi, \xi)$$

for every x in A.

To construct π, one first considers the left regular representation π_0, which represents A as an algebra of linear transformations. Specifically, for each x in A, the linear transformation $\pi_0(x)$ is defined on A by $\pi_0(x): y \to xy$. Clearly π_0 preserves the ring operations of A and respects scalar multiplication. Moreover, $[x, y] = f(y^*x)$ defines a positive semidefinite sesquilinear form on $A \times A$ which satisfies

$$[\pi_0(x)y, z] = [y, \pi_0(x^*)z] \tag{1.6.4}$$

for all x, y, z in A. For z fixed, the usual manipulations with the Schwarz inequality show that the condition $[z, z] = 0$ is equivalent to the condition $[x, z] = 0$ for every x in A. Thus the set $N = \{z \in A : f(z^*z) = 0\}$ is a linear subspace of A. Moreover, 1.6.4 implies that $[y, xz] = [x^*y, z]$, from which it follows that N is a left ideal in A. In particular, N is an invariant subspace for every operator $\pi_0(x)$, $x \in A$.

This allows us to lift each operator $\pi_0(x)$ in the natural way to a linear transformation $\pi(x)$ in the quotient space A/N. Similarly, we can define an inner product $(\cdot, \cdot)$ in A/N by

$$(x + N, y + N) = [x, y].$$

The relation 1.6.4 persists in the form

$$(1.6.5) \qquad (\pi(x)\eta, \zeta) = (\eta, \pi(x^*)\zeta),$$

$x \in A$, $\eta, \zeta \in A/N$. Now suppose, for the moment, that we already know each operator $\pi(x)$ is bounded in the Hilbert norm on A/N. Then we may complete A/N to obtain a Hilbert space $\mathscr{H}$, and every operator $\pi(x)$ extends uniquely to a bounded operator on $\mathscr{H}$, which we denote by the same symbol $\pi(x)$. The map $x \to \pi(x)$ preserves the algebraic operations of A, and 1.6.5 implies that $\pi(x^*) = \pi(x)^*$. Thus, π is a representation of A. Noting finally that the vector $\xi = e + N$ in $\mathscr{H}$ satisfies $(\pi(x)\xi, \xi) = [xe, e] = f(e^*xe) = f(x)$, the desired representation of f is achieved.

It remains to show that each linear transformation $\pi(x)$ is bounded. Actually, we will show that $\|\pi(x)\eta\| \leqslant \|x\| \cdot \|\eta\|$ for every $\eta \in A/N$, $x \in A$. Since η must have the form $y + N$ for some $y \in A$, and since $\|\pi(x)\eta\|^2 = \|xy + N\|^2 = f((xy)^*xy) = f(y^*x^*xy)$, this reduces to the inequality $f(y^*x^*xy) \leqslant \|x\|^2 f(y^*y)$. To prove the latter, fix y and consider the linear functional $g(z) = f(y^*zy)$. Since f is positive, so is g, and by 1.6.2 we know g has norm $g(e) = f(y^*y)$. It follows that $g(x^*x) \leqslant \|x^*x\| f(y^*y) \leqslant \|x\|^2 f(y^*y)$, and the proof is complete. □

Thus every positive linear functional f can be represented in the form $f(x) = (\pi(x)\xi, \xi)$, where π and ξ are as above. We remark that one may always assume that ξ is a *cyclic vector* for π in the sense that $\pi(A)\xi$ is dense in the underlying space. For if this is not so, simply replace ξ with its projection on the subspace $[\pi(A)\xi]$ and π with the corresponding subrepresentation; these new quantities serve equally to represent f, and it is very easy to check that they have the desired property.

Recall that a positive linear functional f is called a *state* if it is normalized so that $f(e) = 1$. Evidently, the states form a convex set of linear functionals on A, and an extreme point of this convex set is called a *pure state*. Thus, a state f is pure iff for any two states g_1 and g_2 and every real number t, $0 < t < 1$, the condition $f = tg_1 + (1 - t)g_2$ implies $g_1 = g_2 = f$. We can now describe an important connection between pure states and irreducible representations.

Theorem 1.6.6. *Let π be a representation of A having a unit cyclic vector ξ, and let f be the state $f(x) = (\pi(x)\xi, \xi)$. Then f is pure if, and only if, π is irreducible.*

PROOF. Assume first that f is pure, and let E be a projection in the commutant of $\pi(A)$. We will prove that $E = 0$ or 1.

Assume, to the contrary, that $E \neq 0$ and $E \neq 1$. Then we claim $E\xi \neq 0$ and $E^\perp\xi \neq 0$. For if, say, $E\xi = 0$, then for every $x \in A$ we have $E\pi(x)\xi = \pi(x)E\xi = 0$, and $E = 0$ follows from the fact that $\pi(A)\xi$ is dense. The same reasoning shows $E^\perp\xi \neq 0$.

It follows that the real number $t = \|E\xi\|^2 = (E\xi, \xi)$ is positive and less than 1. Define linear functionals g_1, g_2 on A by $g_1(x) = t^{-1}(\pi(x)\xi, E\xi)$ and $g_2(x) = (1 - t)^{-1}(\pi(x)\xi, E^\perp\xi)$. Clearly $f = tg_1 + (1 - t)g_2$, and we claim now that each g_i is a state. Indeed, since we can write $(\pi(x)\xi, E\xi) = (\pi(x)\xi, E^2\xi) = (E\pi(x)\xi, E\xi) = (\pi(x)E\xi, E\xi)$, it is apparent that $g_1(x) = t^{-1}(\pi(x)E\xi, E\xi)$ is positive, and satisfies $g_1(e) = t^{-1}\|E\xi\|^2 = 1$. Similarly, g_2 is a state. Since f is an extreme point, we conclude, in particular, that $g_1 = f$ or, what is the same, $(\pi(x)\xi, E\xi) = t(\pi(x)\xi, \xi)$ for every x in A. This implies $(\pi(x)\xi, E\xi - t\xi) = 0$, and since $\pi(A)\xi$ is dense, we see that $E\xi - t\xi = 0$. Since $E - t1$ commutes with $\pi(A)$ and annihilates ξ, we may repeat an argument given before to conclude that $E - t1 = 0$. But this is absurd because E is a projection and $0 < t < 1$.

Conversely, assume π is irreducible, and suppose we are given states g_i and $t \in (0, 1)$ such that $f = tg_1 + (1 - t)g_2$. We will show that $g_1 = f$. To that end, define a sesquilinear form $[\cdot, \cdot]$ on the dense linear manifold $\pi(A)\xi$ as follows:

$$[\pi(x)\xi, \pi(y)\xi] = tg_1(y^*x).$$

Because $0 \leqslant tg_1(x^*x) = f(x^*x) - (1 - t)g_2(x^*x) \leqslant f(x^*x)$, it follows that $0 \leqslant [\pi(x)\xi, \pi(x)\xi] \leqslant \|\pi(x)\xi\|^2$, and in particular $[\cdot, \cdot]$ is bounded. By a familiar lemma of Riesz, there is an operator H on the underlying Hilbert space satisfying $0 \leqslant H \leqslant 1$, and $[\eta, \zeta] = (\eta, H\zeta)$ for all η, ζ in $\pi(A)\xi$. Taking $\eta = \pi(y)\xi$ and $\zeta = \pi(z)\xi$ we obtain $tg_1(z^*y) = (\pi(y)\xi, H\pi(z)\xi)$.

We claim now that H commutes with $\pi(A)$. Since $\pi(A)\xi$ is dense, this amounts to showing that

$$(\pi(y)\xi, H\pi(x)\pi(z)\xi) = (\pi(y)\xi, \pi(x)H\pi(z)\xi)$$

for every x, y, z in A. But the left side is $(\pi(y)\xi, H\pi(xz)\xi) = tg_1((xz)^*y)$, and the right side is $(\pi(x)^*\pi(y)\xi, H\pi(z)\xi) = (\pi(x^*y)\xi, H\pi(z)\xi) = tg_1(z^*x^*y)$, from which the assertion is evident.

Since π is irreducible, the commutant of $\pi(A)$ consists of scalar operators, and we conclude that there is a (real) scalar r such that $H = r1$. Hence, $tg_1(y) = (\pi(y)\xi, H\xi) = r(\pi(y)\xi, \xi) = rf(y)$. Setting $y = e$ and noting that $f(e) = g_1(e) = 1$, we see that $t = r$, and finally we conclude that $g_1 = f$. The proof that $g_2 = f$ is the same. □

EXERCISES

1.6.A. For every unit vector ξ in a Hilbert space $\mathscr{H}$, let ω_ξ be the linear functional on $\mathscr{L}(\mathscr{H})$ defined by

$$\omega_\xi(T) = (T\xi, \xi).$$

a. Show that each ω_ξ is a pure state.

b. Show that the representations of $\mathscr{L}(\mathscr{H})$ associated to the ω_ξ's via the GNS construction are all equivalent to the identity representation.

1.6.B. Let π, σ be two representations of a C^*-algebra A having respective cyclic vectors ξ, η. Assuming that $(\pi(x)\xi, \xi) = (\sigma(x)\eta, \eta)$ for each $x \in A$, show that π and σ are equivalent.

1.6.C. Let $\mathscr{A}$ be the C^*-algebra $C^*(S)$ generated by the unilateral shift (see 1.4.D and 1.5.D).

a. Find all irreducible representations of $\mathscr{A}$ (up to unitary equivalence).

b. Let ρ be a pure state of $\mathscr{A}$ which does not vanish on $\mathscr{C}(\mathscr{H})$. Show that there is a unit vector ξ in $\mathscr{H}$ such that $\rho(T) = (T\xi, \xi)$, $T \in \mathscr{A}$.

c. Find all the remaining pure states of $\mathscr{A}$.

1.6.D. Let A be the algebra of all polynomials in a single real variable having complex coefficients. Give A the norm

$$\|f\| = \sup_{0 \leqslant t \leqslant 1} |f(t)|$$

and the involution $f^*(t) = \overline{f(t)}$.

a. Show that A has all the properties of a commutative C^*-algebra with identity except completeness.

b. Show that $\rho(f) = f(2)$ defines a linear functional on A which satisfies $\rho(f^*f) \geqslant 0$ for all f, but which is not continuous.

1.6.E. Let A be a C^*-algebra without unit and let f be a linear functional on A satisfying $f(x^*x) \geqslant 0$ for all x. Let A_1 be the C^*-algebra obtained from A be adjoining a unit e, let M be a positive constant, and define $\tilde{f}: A_1 \to C$ by

$$f(\lambda e + x) = \lambda M + f(x),$$

$\lambda \in C$, $x \in A$. Prove that in order for $\tilde{f}$ to be positive it is necessary for f to be bounded and that M should satisfy $M \geqslant \|f\|$ (note that 1.6.2 does not apply to nonunital C^*-algebras).

1.7. The Existence of Representations

The purpose of this section is to show that an abstract C^*-algebra has "sufficiently many" irreducible representations, and to present a famous theorem of Gelfand and Naimark on the concrete representation of such algebras. These results rest on the fact that self-adjoint elements of the form z^*z must have nonnegative spectra. This basic property of C^*-algebras is proved in 1.7.1, after one or two preliminaries.

As in the preceding section, A will denote an abstract C^*-algebra with unit e. If z is an element of A, we will use the notation $\mathrm{sp}(z) \geqslant 0$ to express the fact that the spectrum of z is contained in the nonnegative real axis. $\mathrm{sp}(z) \leqslant 0$ has the obvious meaning. We first want to point out that the condition $\mathrm{sp}(z^*z) \geqslant 0$ is more or less obvious when $z = z^*$. Indeed, we know from Section 1.1 that $\mathrm{sp}(z)$ is real, and so the spectral mapping theorem implies that $\mathrm{sp}(z^2) = \{\lambda^2 : \lambda \in \mathrm{sp}(z)\} \geqslant 0$.

Lemma 1. *If x and y are self-adjoint elements of A with $\mathrm{sp}(x) \geqslant 0$ and $\mathrm{sp}(y) \geqslant 0$, then $\mathrm{sp}(x + y) \geqslant 0$.*

PROOF. By multiplying the sum $x + y$ by a sufficiently small positive scalar, if necessary, we may assume $\|x\| \leqslant 1$ and $\|y\| \leqslant 1$. Now since the spectral radius does not exceed the norm, the spectrum of x must be contained in the real interval $[0, +1]$. Thus, $\mathrm{sp}(e - x) \subseteq [0, +1]$. Since $e - x$ is self-adjoint, its norm agrees with its spectral radius, and we have $\|e - x\| = r(e - x) \leqslant 1$. Similarly, $\|e - y\| \leqslant 1$. It follows from the triangle inequality that $e - \frac{1}{2}(x + y) = \frac{1}{2}(e - x) + \frac{1}{2}(e - y)$ has norm at most 1, and so the spectrum of the self-adjoint element $\frac{1}{2}(x + y)$ must lie in the interval $[0, +2]$. The desired conclusion is now immediate. □

In the following lemma, we shall have to make use of the fact that if x and y are elements of an arbitrary Banach algebra with unit, then the nonzero points in $\mathrm{sp}(xy)$ and $\mathrm{sp}(yx)$ are the same (Exercise 1.7.A).

Lemma 2. *Let $z \in A$ be such that $\mathrm{sp}(z^*z) \leqslant 0$. Then $z = 0$.*

PROOF. Assume that $\mathrm{sp}(z^*z) \leqslant 0$. Then by the preceding remark we also have $\mathrm{sp}(zz^*) \leqslant 0$. Lemma 1, applied to the elements $-z^*z$ and $-zz^*$, implies that $\mathrm{sp}(z^*z + zz^*) \leqslant 0$.

On the other hand, let $z = x + iy$ be the Cartesian decomposition of z, where x and y are self-adjoint. A direct computation shows that $z^*z + zz^* = x^2 + y^2$. We have already remarked that $\mathrm{sp}(x^2) \geqslant 0$ and $\mathrm{sp}(y^2) \geqslant 0$; so by Lemma 1 once again, we conclude that $\mathrm{sp}(z^*z + zz^*) = \mathrm{sp}(x^2 + y^2) \geqslant 0$.

Combining these two assertions, we see that the self-adjoint element $z^*z + zz^* = x^2 + y^2$ has spectrum $\{0\}$, and hence $x^2 + y^2 = 0$. Note that this implies $x = y = 0$. For $\mathrm{sp}(x^2) \geqslant 0$, and since $x^2 = -y^2$, we also have $\mathrm{sp}(x^2) \leqslant 0$. As before, we conclude that $x^2 = 0$, and therefore $x = 0$. Similarly, $y = 0$. That proves $z = x + iy = 0$. □

Theorem 1.7.1. *For every z in A, one has $\mathrm{sp}(z^*z) \geqslant 0$.*

PROOF. We claim first that there exist self-adjoint elements u, v in A for which $uv = vu = 0$ and $z^*z = u^2 - v^2$. Indeed, if we consider the real-valued continuous functions f, g defined on $\mathbb{R}$ by

$$f(t) = \begin{cases} \sqrt{t} & t \geqslant 0 \\ 0 & t < 0, \end{cases}$$

and

$$g(t) = \begin{cases} 0 & t \geqslant 0 \\ \sqrt{-t} & t < 0, \end{cases}$$

then $fg = 0$ and $f(t)^2 - g(t)^2 = t$. It follows that $u = f(z^*z)$ and $v = g(z^*z)$ have the asserted properties.

Thus, we can write $z^*z = u^2 - v^2$ as above. Because $uv = 0$, it follows that $vz^*zv = vu^2v - v^4 = -v^4$; and since $\mathrm{sp}(-v^4) = -\mathrm{sp}(v^4) \leqslant 0$, we conclude that the element $w = zv$ satisfies $\mathrm{sp}(w^*w) \leqslant 0$. By Lemma 2, $-v^4 = w^*w = 0$, hence $v^4 = 0$, and so $v = 0$ because v is self-adjoint. All of this shows that $z^*z = u^2$ is the square of a *self-adjoint* element of A, and the conclusion $\mathrm{sp}(z^*z) \geqslant 0$ now follows from the remark preceding Lemma 1. □

We have seen in the preceding section that a state must have norm 1. The following provides a useful converse to this assertion.

Corollary. *Let f be a bounded linear functional on A such that $\|f\| = f(e) = 1$. Then f is a state.*

PROOF. We have to show that $f(z^*z) \geqslant 0$ for every $z \in A$. By 1.7.1, it suffices to show that $f(y) \geqslant 0$ for every self-adjoint element $y \in A$ having nonnegative spectrum. Actually, we will prove this: for every $y \in A$ satisfying $y^*y = yy^*$, $f(y)$ belongs to the closed convex hull K of $\mathrm{sp}(y)$. For that, suppose, to the contrary, that $f(y)$ does not belong to K. Since K is the intersection of all closed discs which contain $\mathrm{sp}(y)$, we infer that there is a complex number a and an $R > 0$ such that $\mathrm{sp}(y)$ is contained in the disc $\{\lambda \in \mathbb{C} : |\lambda - a| \leqslant R\}$, but $|f(y) - a| > R$. The first assertion implies that the spectral radius of $y - ae$ does not exceed R, and since $y - ae$ generates a *commutative* C^*-subalgebra, we know from 1.1.1 that $\|y - ae\| = r(y - ae) \leqslant R$. On the other hand, $|f(y - ae)| = |f(y) - a| > R$, and these two assertions contradict the fact that $\|f\| = 1$. □

We are now in a position to prove a key result on the existence of states.

Theorem 1.7.2. *Let x be a self-adjoint element of A. Then there is a pure state f such that $|f(x)| = \|x\|$.*

PROOF. We claim first that there is a state f with the property $|f(x)| = \|x\|$. Consider the C^*-subalgebra B generated by x and e. Since B is commutative, it follows from 1.1.1 that there is a complex homomorphism ω of B such that $|\omega(x)| = \|x\|$ (simply choose ω corresponding to some point of the maximal ideal space of B at which the Gelfand transform of x achieves its maximum modulus). We already know that such an ω satisfies $\|\omega\| = \omega(e) = 1$. By the Hahn–Banach theorem, there is an extension f of ω to A which satisfies $\|f\| = 1$. The preceding corollary implies that f is a state, and by construction we have $|f(x)| = |\omega(x)| = \|x\|$.

Now fix such a state f, and let $\sum$ be the set of all states g of A satisfying $g(x) = f(x)$. $\sum$ is clearly closed in the weak *-topology of the unit ball of the dual of A, and therefore, by Alaoglu's theorem, it is compact. $\sum$ is also convex, and the preceding paragraph shows $\sum$ is not void. Thus, the Krein–Milman theorem implies that $\sum$ has an extreme point g.

To complete the proof, we have to show that g is an extreme state. But if g_1 and g_2 are states and $0 < t < 1$ are such that $g = tg_1 + (1 - t)g_2$, then notice that g_1 and g_2 must belong to $\sum$. For the conditions $|g_i(x)| \leqslant \|x\| = |f(x)|$, together with $tg_1(x) + (1 - t)g_2(x) = g(x) = f(x)$, imply that $g_1(x) = g_2(x) = f(x)$, i.e., $g_i \in \sum$. Because g is extreme in $\sum$, we have the desired conclusion $g_1 = g_2 = g$. □

Corollary. *For every nonzero element z in A, there is an irreducible representation π of A on a Hilbert space $\mathscr{H}$ and a unit vector ξ in $\mathscr{H}$ such that*

$$\|\pi(z)\xi\| = \|z\| > 0.$$

Proof. Taking $x = z^*z$ in the preceding theorem, we obtain a pure state f of A such that $f(z^*z) = \|z^*z\| = \|z\|^2$. Let π and ξ be the representation and unit vector associated with f via the GNS construction (1.6.3). Then

$$\begin{aligned}\|\pi(z)\xi\|^2 &= (\pi(z)\xi, \pi(z)\xi) = (\pi(z^*z)\xi, \xi)\\ &= f(z^*z) = \|z\|^2,\end{aligned}$$

and hence $\|\pi(z)\xi\| = \|z\|$. By 1.6.6, π is irreducible, and the proof is complete. □

This corollary can be used to prove that every locally compact group has "sufficiently many" irreducible unitary representations. Unfortunately, an adequate discussion of this important application would take us too far afield, and instead we refer the reader to [6].

We can now deduce the theorem of Gelfand and Naimark mentioned at the beginning of the section.

Theorem 1.7.3. Gelfand–Naimark theorem. *Every abstract C^*-algebra with identity is isometrically *-isomorphic to a C^*-algebra of operators.*

Proof. We have to exhibit an isometric representation π of A. By the preceding corollary we may choose, for each $x \neq 0$ in A, a representation π_x such that $\|\pi_x(x)\| = \|x\|$. Now simply let π be the direct sum of all these π_x's. □

This theorem and the corollary of 1.7.2 are valid for C^*-algebras without identities, and in fact can be deduced from what we have done by adjoining an identity (see Exercise 1.7.C).

EXERCISES

1.7.A. Let A be a Banach algebra with unit e, and let x, y belong to A. Show that $e - xy$ is invertible if and only if $e - yx$ is invertible. Deduce that $\mathrm{sp}(xy)$ and $\mathrm{sp}(yx)$ agree except possibly for the point 0.

1.7.B. Let A be a C^*-algebra with unit e and let x be a self-adjoint element of A satisfying $\|x\| \leqslant 1$. Show that $\mathrm{sp}(x) \geqslant 0$ if and only if $\|e - x\| \leqslant 1$.

1.7.C. Generalize the corollary of 1.7.2 and Theorem 1.7.3 to C^*-algebras without identities.

1.7.D. Show that the C^*-algebra $\mathscr{L}(\mathscr{H})$ has pure states other than the states ω_ξ, $\xi \in H$, of Exercise 1.6.A.

1.8. Order and Approximate Units

Approximate units have already been discussed rather briefly in Section 1.3. We are now going to indicate how approximate units can be constructed in general C^*-algebras, and we will complete the discussion of positive linear functionals. While these results are important for some applications of C^*-algebra theory we will not require them here, and the reader may omit this section without essential loss.

Let A be a C^*-algebra, with or without unit. Given a self-adjoint element x in A, it is customary to write $x \geqslant 0$ if the spectrum of x is nonnegative (if A has no unit then it is understood that $\mathrm{sp}(x)$ is computed in the C^*-algebra obtained by adjoining a unit to A). Such elements are called *positive*. If x and y are two self-adjoint elements, then $x \leqslant y$ (or $y \geqslant x$) will mean that $y - x$ is positive. To say that this relation is transitive means that the sum of two positive elements is positive, a fact that has already been established in Lemma 1 of the preceding section. For the same reason, we see that if $x_i \leqslant y_i$ for $i = 1, 2, \ldots, n$, then $x_1 + \cdots + x_n \leqslant y_1 + \cdots + y_n$. Finally, notice that $x \leqslant y \leqslant x$ implies $y = x$ (because $y - x$ is then a self-adjoint element of A having spectral radius zero).

Thus $\leqslant$ is a partial ordering of the set of self-adjoint elements of A. If X is a locally compact Hausdorff space and we take A to be the concrete C^*-algebra $C_0(X)$ of all complex-valued continuous functions vanishing at infinity, then $\leqslant$ reduces to the usual partial order $f \leqslant g$ iff $f(p) \leqslant g(p)$ for every p in X. Carrying this a bit farther, the reader can easily see that the Gelfand map of an abstract commutative C^*-algebra A is also an *order isomorphism* in the sense that $x \leqslant y$ in A iff $\hat{x} \leqslant \hat{y}$ as functions in $C_0(\hat{A})$.

We remind the reader that some care must be exercised when utilizing the functional calculus in C^*-algebras A which have no identity: one may only apply continuous functions which vanish at the origin to self-adjoint elements of A (see Exercise 1.1.I). After this qualification, the preceding remarks show that the functional calculus preserves order even in the case of noncommutative C^*-algebras. For instance, if f and g are two (real-valued) continuous functions of a real variable such that $f(t) \leqslant g(t)$ for every real t, then $f(x) \leqslant g(x)$ for every self-adjoint element x in A.

There are two algebraic characterizations of this partial ordering which are frequently useful, and are summarized as follows.

Proposition 1.8.1. *Let x be a self-adjoint element of A. Then the following are equivalent:*

(i) $x \geqslant 0$.
(ii) *There is an element $y = y^*$ in A such that $x = y^2$.*
(iii) *There is an element z in A such that $x = z^*z$.*

PROOF. That (ii) implies (iii) is a triviality, (iii) implies (i) is 1.7.1, and (i) implies (ii) is a simple exercise with the functional calculus which we leave for the reader. □

This result implies an invariance property of the partial order which is often useful: *if $x \leqslant y$ then $z^*xz \leqslant z^*yz$ for every element z in A*. To see why, use (ii) to find $u = u^*$ in A with $y - x = u^2$ and note that $z^*yz - z^*xz$ has the form w^*w with $w = uz$.

An *approximate unit* (or *approximate identity*) for a C^*-algebra A is a net e_λ of self-adjoint elements of A satisfying

(i) $e_\lambda \geqslant 0$, $\|e_\lambda\| \leqslant 1$;
(ii) $\lambda \leqslant \mu$ implies $e_\lambda \leqslant e_\mu$;
(iii) $\lim_\lambda \|xe_\lambda - x\| = \lim_\lambda \|e_\lambda x - x\| = 0$, for every x in A.

Before giving the construction of approximate units, we need a preliminary result.

Lemma. *Assume A has a unit and let x, y be invertible positive elements of A satisfying $x \leqslant y$. Then $y^{-1} \leqslant x^{-1}$.*

PROOF. Notice first that for every $z \in A$ the condition $z^*z \leqslant e$ (e denoting the unit of A) is equivalent to $\|z\| \leqslant 1$. For example, if $\|z\| \leqslant 1$ then $\|z^*z\| \leqslant 1$, and by 1.7.1 we see that $\operatorname{sp}(z^*z)$ is contained in the unit interval $[0, 1]$. Hence $\operatorname{sp}(e - z^*z) \geqslant 0$. The converse is similar, and we omit the proof.

Since both x and y are positive, we may find positive square roots for each of them (using the functional calculus), which are denoted by $x^{1/2}$ and $y^{1/2}$. Now multiply $x \leqslant y$ on left and right by $y^{-1/2}$ to obtain $y^{-1/2}xy^{-1/2} \leqslant e$. The preceding paragraph, applied to $z = x^{1/2}y^{-1/2}$, implies $\|x^{1/2}y^{-1/2}\| \leqslant 1$, hence $\|y^{-1/2}x^{1/2}\| \leqslant 1$ (by taking the adjoint), and so $zz^* \leqslant e$. That is simply $x^{1/2}y^{-1}x^{1/2} \leqslant e$, and now the conclusion $y^{-1} \leqslant x^{-1}$ follows after another double multiplication by $x^{-1/2}$. □

Theorem 1.8.2. *Every C^*-algebra has an approximate unit.*

PROOF. We may as well assume that the given algebra A has no unit, and let A_1 be the C^*-algebra obtained from A by adjoining one.

Let Λ be the directed set of all finite sets of self-adjoint elements of A (ordered in the usual way by set inclusion), and for each integer $n \geqslant 1$ let f_n be the function defined on the interval $[0, +\infty)$ by

$$f_n(t) = nt(1 + nt)^{-1}.$$

Because f_n is continuous and vanishes at the origin, it operates on the set of all *positive* elements of A. So for each $\lambda \in \Lambda$, say $\lambda = \{x_1, \ldots, x_n\}$, we may define a self-adjoint element e_λ in A by

$$e_\lambda = f_n(x_1^2 + \cdots + x_n^2).$$

Because $x_1^2 + \cdots + x_n^2 \geqslant 0$ and f_n maps $[0, +\infty)$ into the unit interval, we have $\text{sp}(e_\lambda) \subseteq [0, 1]$, so that property (i) in the definition of approximate units is satisfied.

For (ii), let $\lambda \leqslant \mu$, say $\lambda = \{x_1, \ldots, x_m\}$ and $\mu = \{x_1, \ldots, x_n\}$ where $m \leqslant n$. To prove $e_\lambda \leqslant e_\mu$ it suffices to show that $e - e_\mu \leqslant e - e_\lambda$ (where both sides of the latter inequality are interpreted as elements of A_1). But since $1 - f_k(t) = (1 + kt)^{-1}$, the latter inequality is simply

$$[e + n(x_1^2 + \cdots + x_n^2)]^{-1} \leqslant [e + m(x_1^2 + \cdots + x_m^2)]^{-1},$$

which follows from the lemma because

$$e + m(x_1^2 + \cdots + x_m^2) \leqslant e + n(x_1^2 + \cdots + x_n^2).$$

To prove (iii), it suffices to show that $\|x - xe_\lambda\| \to 0$ for every self-adjoint x in A; for this condition implies $\|x - e_\lambda x\| \to 0$ (simply by taking adjoints), and so (iii) follows from these two conditions because every element of A can be decomposed into the form $x_1 + ix_2$ with $x_i = x_i^*$.

So fix $x = x^*$ in A, let m be a positive integer, and let $\lambda = \{x_1, \ldots, x_n\}$ be any finite set of self-adjoint elements which contains x and has at least m elements. Then we claim that $\|x - xe_\lambda\|^2 \leqslant 1/4m$. Indeed, since $x^2 \leqslant x_1^2 + \cdots + x_n^2$ we have

$$(e - e_\lambda)x^2(e - e_\lambda) \leqslant (e - e_\lambda)(x_1^2 + \cdots + x_n^2)(e - e_\lambda).$$

Now the right side has the form $g_n(x_1^2 + \cdots + x_n^2)$, where $g_n(t) = (1 - f_n(t))\cdot t(1 - f_n(t)) = t(1 + nt)^{-2}$. Since g_n is bounded above on $[0, +\infty)$ by $1/4n$, we conclude that $(e - e_\lambda)x^2(e - e_\lambda) \leqslant (4n)^{-1}e$. This asserts that the element $z = x - xe_\lambda = x(e - e_\lambda)$ satisfies $z^*z \leqslant (4n)^{-1}e$, and an argument already given in the proof of the lemma shows that $\|z\|^2 \leqslant (4n)^{-1} \leqslant (4m)^{-1}$. (iii) now follows because m was arbitrary. □

It is significant that if A is separable, then the net e_λ can be chosen to be a *sequence* $e_1, e_2, \ldots$. For if $x_1, \ldots$ is a sequence dense in the set of self-adjoint elements of A then, in the notation of the preceding proof, one may define e_n by

$$e_n = e_{\{x_1, \ldots, x_n\}}.$$

Properties (i) and (ii) are clearly satisfied for this new sequence, and the proof of (iii) is a minor variation of the above which is left for the reader.

It was shown in Section 1.6 that a positive linear functional on a C^*-algebra A with unit e is bounded, and in fact achieves its norm at e. Suppose now that f is a positive linear functional on a nonunital C^*-algebra A. It is tempting to try to prove f is bounded by writing down a positive extension $\tilde{f}$ of f to the C^*-algebra A_1 obtained by adjoining a unit e to A, and then utilizing the preceding result. However, $\tilde{f}$ must have the form $\tilde{f}(te + x) = tM + f(x)$ ($t \in \mathbb{R}$, $x \in A$), where M is some appropriately chosen positive constant, and it is not hard to see that in order for $\tilde{f}$ to be positive it is necessary to choose M at least as large as

$$\sup\{f(x): x = x^*, \operatorname{sp}(x) \subseteq [-1, +1]\} = \sup\{f(x): x = x^*, \|x\| \leqslant 1\}.$$

So it is necessary that the latter supremum be finite, and that requires f to be bounded already.

What is required is the following result, which generalizes Proposition 1.6.2.

Theorem 1.8.3. *Let f be a positive linear functional on a C^*-algebra A and let e_λ be an approximate unit for A. Then f is bounded, and $\|f\| = \lim_\lambda f(e_\lambda)$.*

PROOF. We first claim that f is bounded on the positive part B^+ of the unit ball of A (here, of course, B^+ denotes the set of self-adjoint elements x satisfying $x \geqslant 0$ and $\|x\| \leqslant 1$). Notice that $f(x) \geqslant 0$ for every x in B^+, so if f is unbounded on B^+ then we may find a sequence $x_n \in B^+$ such that $f(x_n) > 1/n$. Consider the infinite series $\sum n^{-2} x_n$. This converges to a self-adjoint element $x \in A$, and since each of the partial sums is positive it follows that $x \geqslant 0$ (see Exercise 1.8.B). Similarly, letting y_k denote the k-th partial sum of the series, we have $x \geqslant y_k$ and so $f(x) \geqslant f(y_k)$. This implies that $f(x) \geqslant \sum n^{-2} f(x_n) \geqslant \sum n^{-1} = +\infty$, a contradiction.

So let $M = \sup\{f(x): x \in B^+\}$. Then for each $x = x^* \in A$ with $\|x\| \leqslant 1$ we can write $x = x_1 - x_2$ with $x_i \in B^+$ (Exercise 1.3.C.), so that $|f(x)| \leqslant 2M$. For an arbitrary $z \in A$ with $\|z\| \leqslant 1$, write $z = y_1 + iy_2$ where $y_i = y_i^*$ and $\|y_i\| \leqslant 1$, and note that $|f(z)| \leqslant |f(y_1)| + |f(y_2)| \leqslant 4M$. That shows f is bounded.

Now since $\lambda \leqslant \mu$ implies $e_\lambda \leqslant e_\mu$ and f is positive, $f(e_\lambda)$ is a (bounded) increasing net; hence $\lim f(e_\lambda)$ exists and is at most $\|f\|$. For the opposite inequality, fix $x \in A$, $\|x\| \leqslant 1$. Then $xe_\lambda \to x$ as $\lambda\uparrow$, so by the Schwarz inequality 1.6.1 (which does *not* depend on the existence of a unit) we have

$$|f(x)|^2 = \lim_\lambda |f(xe_\lambda)|^2 \leqslant \limsup_\lambda f(e_\lambda^2) f(x^*x).$$

Since $\|x^*x\| \leqslant 1$ we have $f(x^*x) \leqslant \|f\|$, and since $e_\lambda^2 \leqslant e_\lambda$ (because of the fact that $t^2 \leqslant t$ over the unit interval and the order properties of the functional calculus) it follows that $f(e_\lambda^2) \leqslant f(e_\lambda)$. So the displayed inequality implies that $|f(x)|^2 \leqslant \|f\| \lim_\lambda f(e_\lambda)$ from which the desired assertion $\|f\| \leqslant \lim_\lambda f(e_\lambda)$ is immediate. □

We have already indicated at the end of Section 1.7 how the results of that section could be derived for nonunital C^*-algebras. With the help of the preceding theorem it is now a simple matter to imitate the GNS construction and the main results of Section 1.6 for such C^*-algebras. The explicit details are left for the reader.

Exercises

1.8.A. Let x, y be self-adjoint elements of a C^*-algebra A satisfying $0 \leqslant x \leqslant y$. Prove that $\|x\| \leqslant \|y\|$.

1.8.B. Let A be a C^*-algebra and let $P = \{x \in A : x \geqslant 0\}$. Show that P is a closed cone in A, dual to the cone of all positive linear functionals on A [*Hint*: see 1.7.B.]

1.8.C. Let J be an ideal in a C^*-algebra A, let $\{e_\lambda\}$ be an approximate unit for J, and let $x \to \dot{x}$ be the projection of A on A/J. Prove that, for each element x in A,

$$\lim_\lambda \|x - xe_\lambda\| = \|\dot{x}\|.$$

1.8.D. Let π be a nondegenerate representation of a C^*-algebra A on a Hilbert space $\mathscr{H}$, and let $\{e_\lambda\}$ be an approximate identity for A. Show that the net of operators $\pi(e_\lambda)$ converges in the strong operator topology to the identity operator on $\mathscr{H}$.

2 Multiplicity Theory

The objects in a number of mathematical categories (groups, C^*-algebras, σ-fields of sets are but three examples) can be represented in a variety of ways by means of operators on Hilbert spaces. For example, the permutation group on n letters acts in an obvious way to permute the elements of an orthonormal basis for an n-dimensional Hilbert space, and this gives rise to a representation of that group as a group of unitary operators. Multiplicity theory has to do with the classification of these representations. This classification problem amounts to finding, for a given object, a suitable collection of representation invariants which will allow one to determine when any two of its representations are, or are not, geometrically the "same"—that is, unitarily equivalent. It goes without saying that in order to classify representations one first has to find them (all of them), so an adequate solution to this vague problem should contain a procedure for *constructing* all possible representations of the object (more precisely, at least one representative from every unitary equivalence class) in usable and concrete terms.

Unfortunately, even for the case of separable C^*-algebras, such a detailed and general classification appears today to be hopelessly out of reach. If one restricts attention to an appropriate subcategory however, namely C^*-algebras having only type I representations, then a quite satisfactory theory is available, and that is the subject of this chapter and Chapter 4. These results contain the Hahn–Hellinger classification of self-adjoint operators, and more generally they give useful information about the structure of operators which generate GCR C^*-algebras.

The work behind this classification divides naturally into two distinct parts. The first step is to analyze the set of operators in the range of a given representation, and is very general in the sense that the analysis for unitary representations of groups is identical with the analysis for representations of C^*-algebras, and so on. The work amounts to a reduction of the problem for general representations to that for "multiplicity-free" representations, and is taken up in 2.1 below. The second step is to classify the multiplicity-free representations of a given object in terms of a suitable set of invariants. The details here vary from one category to another, but generally the invariants turn out to be measures (more precisely, measure classes) defined on a canonical σ-field of subsets of a certain set (the dual object) associated with the given object. This part of the analysis is carried out for commutative C^*-algebras in Section 2.2, and for the larger class of GCR algebras in Chapter 4.

2.1. From Type I to Multiplicity-Free

For definiteness, we shall consider representations of a C^*-algebra A, fixed throughout this section, and in this section all nonzero representations of A are assumed to be nondegenerate. Let us first recall one or two simple facts about representations of A. If σ is a subrepresentation of a representation π of A and $\mathfrak{M}$ is the range space of σ, then since $\pi(A)$ is a self-adjoint family of operators $\mathfrak{M}^\perp$ is also $\pi(A)$-invariant and thus defines a subrepresentation $\sigma^\perp$. σ and $\sigma^\perp$ are orthogonal in an obvious sense and we have $\sigma + \sigma^\perp = \pi$. More generally, if $\{\sigma_i\}$ is any family of mutually orthogonal subrepresentations of π then $\sum_i \sigma_i$ has an obvious meaning and defines another subrepresentation of π. Given two such families $\{\sigma_i\}$ and $\{\sigma_i'\}$ for which σ_i is equivalent to σ_i' for all i, then a simple "piecing together" argument shows that the subrepresentations $\sum_i \sigma_i$ and $\sum_i \sigma_i'$ are equivalent. The notation $\sigma_1 \perp \sigma_2$ means that σ_1 and σ_2 are orthogonal subrepresentations of π. Finally, it is a simple consequence of the definitions that two (not necessarily orthogonal) subrepresentations σ_1, σ_2 of π are equivalent if, and only if, there is a partial isometry in the commutant $\pi(A)'$ with initial and final spaces equal, respectively, to the range spaces of σ_1 and σ_2.

The term "multiplicity" is supposed to suggest the presence of multiple copies of the same representation. Thus, if σ is a given representation of A on a Hilbert space $\mathscr{H}$ then $\pi(x) = \sigma(x) \oplus \sigma(x) \oplus \sigma(x)$, $x \in A$, defines a representation of A on $\mathscr{H} \oplus \mathscr{H} \oplus \mathscr{H}$ which apparently *has* multiplicity. It would be too naive to say that π has multiplicity 3 however, since, for example, σ itself might admit such a multiple decomposition. Our starting point, then, will be to isolate those representations which are multiplicity-free in the sense that this kind of multiple decomposition cannot occur.

Definition 2.1.1. A representation π of A is said to be *multiplicity-free* if π does not have two nonzero orthogonal equivalent subrepresentations.

In other words, π is multiplicity-free iff the only representation σ of A for which $\sigma \oplus \sigma$ is equivalent to a subrepresentation of π is the zero representation. It is apparent from the definition that *every subrepresentation of a* *multiplicity-free representation is multiplicity-free.* We shall make repeated use of this observation. Usually, one finds a different definition than 2.1.1, namely the following: π is multiplicity-free if the von Neumann algebra $\pi(A)'$ is abelian. To see, briefly, that the two are equivalent, suppose first that π has two nonzero orthogonal equivalent subrepresentations σ_1 and σ_2. By the preceding remarks there is a partial isometry $U \in \pi(A)'$ having initial and final spaces respectively the range spaces of σ_1 and σ_2. Thus U^*U and UU^* are orthogonal nonzero projections and in particular U and U^* do not commute, proving that $\pi(A)'$ is not abelian. Conversely, if $\pi(A)'$ is not abelian then there is a self-adjoint operator $T \in \pi(A)'$ which does not commute with $\pi(A)'$; an application of the spectral theorem yields a projection E with the same properties. Thus the range of E cannot be invariant under every operator in $\pi(A)'$ and so there exists S in $\pi(A)'$ such that $(I - E)SE \neq 0$. Now the initial and final spaces of $(I - E)SE$ are clearly orthogonal, and so the polar decomposition applied to this operator yields a nonzero partial isometry $V \in \pi(A)'$ such that V^*V and VV^* are orthogonal projections. Thus, the ranges of V^*V and VV^* give rise to two nonzero orthogonal equivalent subrepresentations of π, proving our assertion.

From here on in this section, we assume all representations act on *separable* Hilbert spaces. The assumption is not essential for this part of the theory, but it allows the development to run more smoothly, and moreover the overwhelming majority of representations that arise in practice are on separable spaces. Later on, in Chapter 4, this assumption will become more essential.

We now make the following tentative definition.

Definition 2.1.2. Let π be a representation of A and let n be a cardinal, $1 \leqslant n \leqslant \aleph_0$. π is said to have *multiplicity* n if there is a set of n mutually orthogonal, mutually equivalent, multiplicity-free subrepresentations $\{\sigma_j\}$ of π such that $\pi = \sum_j \sigma_j$.

According to the following proposition, the multiplicity of a representation is well-defined. First, some notation. If π is a representation of A on $\mathscr{H}$ and n is a cardinal, $1 \leqslant n \leqslant \aleph_0$, define $n \cdot \pi$ to be the representation $n \cdot \pi(x) = \pi(x) \oplus \pi(x) \oplus \cdots$ on $\mathscr{H} \oplus \mathscr{H} \oplus \cdots$, where the number of summands is n. Observe that a representation π has multiplicity n iff there is a multiplicity-free representation σ such that $\pi \sim n \cdot \sigma$.

Proposition 2.1.3. *Let $\{\mu_i : i \in I\}$ and $\{\nu_j : j \in J\}$ be two countable families of subrepresentations of π, each satisfying the three conditions of 2.1.2, such that $\pi = \sum \mu_i = \sum \nu_j$. Then* card I = card J.

PROOF. Let m = card I, n = card J, and suppose, to the contrary, that $1 \leqslant m < n \leqslant \aleph_0$. Choose $i_0 \in I$ and $j_0 \in J$, and let $\mathscr{M}$ and $\mathscr{N}$ be the respective commutants of $m \cdot \mu_{i_0}(A)$ and $n \cdot \nu_{j_0}(A)$. From the hypothesis it follows that $m \cdot \mu_{i_0}$ and $n \cdot \nu_{j_0}$ are each equivalent to π (cf. the remarks preceding 2.1.1), hence $\mathcal{M}$ and $\mathcal{N}$ are unitarily equivalent to $\pi(A)'$ and they are in particular $\star$-isomorphic as C^*-algebras.

Now as in the proof of 1.2.1 the elements of $\mathcal{M}$ can be realized as $m \times m$ matrices over the abelian von Neumann algebra $\mu_{i_0}(A)'$ and similarly $\mathcal{N}$ consists of all $n \times n$ matrices over $\nu_{j_0}(A)'$ (assume n is finite; the infinite case will be dealt with separately). Note first that $\mathcal{M}$ admits a $\star$-homomorphism onto M_m, the algebra of all $m \times m$ matrices over the complex numbers; simply choose a nonzero homomorphism ω of the commutative C^*-algebra $\mu_{i_0}(A)'$ into $\mathbb{C}$ and define $\psi_\omega : \mathcal{M} \to M_n$ by applying ω element by element to each matrix over $\mu_{i_0}(A)'$. Thus $\mathcal{N}$ too admits a $\star$-homomorphism φ_ω onto M_m. Let $\mathcal{N}_0$ be the subalgebra of $\mathcal{N}$ consisting of all $n \times n$ operator matrices having scalar entries. Then $\mathcal{N}_0$ is clearly $\star$-isomorphic with M_n, and $\varphi_\omega(\mathcal{N}_0) \neq 0$ (since $\mathcal{N}_0$ contains the identity), thus $\varphi_\omega|_{\mathcal{N}_0}$ has trivial kernel because M_n has no nonzero proper ideals. In particular $\varphi_\omega(\mathcal{N}_0)$ is an n^2-dimensional linear subspace of the m^2-dimensional space M_m, contradicting $m < n$.

In the case where n is infinite, the argument can be adapted as follows. Note first that $\mathcal{N}$ contains an isometry U (i.e., $U^*U = I$) which is not unitary (i.e., $UU^* \neq I$). Indeed, letting $\mathcal{H}$ be the range space of ν_{j_0}, then $n \cdot \nu_{j_0} = \nu_{j_0} \oplus \nu_{j_0} \oplus \cdots$ acts on $\mathcal{H} \oplus \mathcal{H} \oplus \cdots$, an infinite direct sum. Thus the "unilateral shift" $U : \xi_1 \oplus \xi_2 \cdots \mapsto 0 \oplus \xi_1 \oplus \xi_2 \oplus \cdots$ ($\xi_i \in \mathcal{H}$) is a nonunitary isometry which commutes with $n \cdot \nu_{j_0}(A)$. Thus $\mathcal{M}$ contains an operator V such that $V^*V = I \neq VV^*$. Now some entry of the nonzero $m \times m$ operator matrix $I - VV^*$ is nonzero, hence there is a complex homomorphism ω of $\mu_{i_0}(A)'$ which is nonzero at that entry. Defining $\psi_\omega : \mathcal{M} \to M_m$ as before and putting $W = \psi_\omega(V) \in M_m$, we see that $W^*W = I$ but $WW^* \neq I$. This, however, is impossible in M_n, a fact easily seen by comparing the trace of $I - W^*W$ to the trace of $I - WW^*$ and noting that a positive matrix with zero trace must be zero. Thus n could not be infinite, and the proof is complete. □

There are two more concepts which are basic to this point of view. The first is Mackey's notion of disjointness; the second is the central cover of a subrepresentation. Two representations π_1 and π_2 are said to be *disjoint* ($\pi_1 \,\mathring{\circ}\, \pi_2$) if no nonzero subrepresentation of π_1 is equivalent to any subrepresentation of π_2. The term has its origin in set theory, and in Section 2.2 we will see that this is precisely a subtle refinement of the usual idea of disjoint sets. It is clear that if $\pi_1 \,\mathring{\circ}\, \pi_2$ and π_1' and π_2' are two other representations equivalent, respectively, to π_1 and π_2, then $\pi_1' \,\mathring{\circ}\, \pi_2'$. Let π be a multiplicity-free representation and let σ_1 and σ_2 be two *orthogonal* subrepresentations of π. Then a brief consideration of the definitions involved shows that $\sigma_1 \,\mathring{\circ}\, \sigma_2$. While this is no longer true if π is not multiplicity-free, we will see presently (2.1.5) that the converse is: *two disjoint subrepresentations of a third are necessarily orthogonal.* There are two characterizations of disjointness which are frequently very useful, and we pause now to present them.

Proposition 2.1.4. *Let π_1 and π_2 be two representations of A. Then the following are equivalent:*

(i) $\pi_1 \circ\!\!\!\!\!\!\smile \pi_2$.
(ii) *The only bounded operator T between the respective Hilbert spaces which satisfies $T\pi_1(x) = \pi_2(x)T$, $x \in A$, is $T = 0$.*
(iii) *The projection $1 \oplus 0$ belongs to the weak closure of $\pi_1 \oplus \pi_2(A)$.*

PROOF. Let $\mathscr{H}_i$ be the space on which π_i acts, $i = 1, 2$.

(i) $\Rightarrow$ (ii). Assume $\pi_1 \circ\!\!\!\!\!\!\smile \pi_2$, and let $T:\mathscr{H}_1 \to \mathscr{H}_2$ satisfy the condition $T\pi_1(x) = \pi_2(x)T$, $x \in A$. By taking adjoints, we conclude that the operator $T^*:\mathscr{H}_2 \to \mathscr{H}_1$ satisfies $T^*\pi_2(x) = \pi_1(x)T^*$, $x \in A$. These two conditions imply that the positive operator T^*T commutes with $\pi_1(A)$ and hence $|T| = (T^*T)^{1/2}$ belongs to $\pi_1(A)'$. So if $T = U|T|$ is the polar decomposition, we see that the partial isometry U implements an equivalence between the subrepresentations of π_1 and π_2 defined by the initial and final spaces of U. Since $\pi_1 \circ\!\!\!\!\!\!\smile \pi_2$ we must therefore have $U = 0$, hence $T = 0$.

(ii) $\Rightarrow$ (iii). Assume (ii). By the double commutant theorem, it suffices to show that $1 \oplus 0$ commutes with every operator T on $\mathscr{H}_1 \oplus \mathscr{H}_2$ which commutes with all operators $\pi_1(x) \oplus \pi_2(x)$, $x \in A$. The obvious computations with 2×2 operator matrices show that such a T has a representation

$$T = \begin{pmatrix} T_{11} & T_{12} \\ T_{21} & T_{22} \end{pmatrix}$$

where, for each i and j, $T_{ij}:\mathscr{H}_j \to \mathscr{H}_i$ satisfies $T_{ij}\pi_j(x) = \pi_i(x)T_{ij}$ for every $x \in A$. Since $\pi_1 \circ\!\!\!\!\!\!\smile \pi_2$, the off-diagonal terms must vanish (by (ii)), so that T is a direct sum $T = T_{11} \oplus T_{22}$. The conclusion is now evident.

The implication (iii) $\Rightarrow$ (i) is a simple variation of what was just done, and will be left for the reader. □

Now let σ be a subrepresentation of π and let $\mathscr{M}$ be the range space of σ. It is easy to see that $[\pi(A)'\mathscr{M}]$ is an invariant subspace for both $\pi(A)'$ and $\pi(A)''$ (and, in particular, for $\pi(A)$), so that its projection belongs to the center of $\pi(A)'$. Let $\bar{\sigma}$ be the subrepresentation of π corresponding to $[\pi(A)'\mathscr{M}]$; $\bar{\sigma}$ is called the *central cover* of σ. If $\sigma = \bar{\sigma}$ (equivalently, if the projection on the range space of σ already belongs to the center of $\pi(A)'$) then σ is called a *central* subrepresentation of π. Central subrepresentations are important because, on the one hand, the projections on their range spaces commute with everything in sight and, on the other, these projections belong to the center of the weak operator closure of $\pi(A)$ (indeed the latter is, by the double commutant theorem, the center of $\pi(A)''$ which in turn coincides with the center of $\pi(A)'$). The relation between disjointness, equivalence, and central covers is expressed in the following result.

Proposition 2.1.5. *Let σ_1 and σ_2 be subrepresentations of a representation π.*

(i) *If σ_1 is equivalent to a subrepresentation of σ_2, then $\bar{\sigma}_1 \leqslant \bar{\sigma}_2$.*
(ii) *If σ_1 is multiplicity-free and $\bar{\sigma}_1 \leqslant \bar{\sigma}_2$, then σ_1 is equivalent to a subrepresentation of σ_2.*
(iii) *If σ_1 and σ_2 are multiplicity-free and $\bar{\sigma}_1 = \bar{\sigma}_2$, then σ_1 and σ_2 are equivalent.*
(iv) *$\sigma_1 \circ \sigma_2$ if, and only if, $\bar{\sigma}_1 \perp \bar{\sigma}_2$.*

PROOF

(i) Let $\mathfrak{M}_i$ be the range space of σ_i, $i = 1, 2$. Then by hypothesis, there is a partial isometry $U \in \pi(A)'$ such that U^*U is the projection on $\mathfrak{M}_1$ and $U\mathfrak{M}_1 \subseteq \mathfrak{M}_2$. Thus, $\pi(A)'\mathfrak{M}_1 = \pi(A)'U^*U\mathfrak{M}_1 \subseteq \pi(A)'U\mathfrak{M}_1 \subseteq \pi(A)'\mathfrak{M}_2$, so that $[\pi(A)'\mathfrak{M}_1] \subseteq [\pi(A)'\mathfrak{M}_2]$, and it now follows from the definition of central cover that $\bar{\sigma}_1 \leqslant \bar{\sigma}_2$.

(iv) Suppose first that $\bar{\sigma}_1$ and $\bar{\sigma}_2$ are not orthogonal, and let E_i be the projection on the range space of σ_i, $i = 1, 2$. Note that there is an operator $T \in \pi(A)'$ such that $E_2TE_1 \neq 0$. For if $E_2\pi(A)'E_1 = \{0\}$ then for all T_1, $T_2 \in \pi(A)'$ and $\xi_i \in E_i\mathscr{H}$ ($\mathscr{H}$ being the underlying space) we have

$$(T_1\xi_1, T_2\xi_2) = (E_2T_2^*T_1E_1\xi_1, \xi_2) = 0,$$

hence $[\pi(A)'E_1\mathscr{H}] \perp [\pi(A)'E_2\mathscr{H}]$, or $\bar{\sigma}_1 \perp \bar{\sigma}_2$, a contradiction. Applying 2.1.4 (ii), we conclude that σ_1 and σ_2 are not disjoint.

Conversely, assume $\bar{\sigma}_1 \perp \bar{\sigma}_2$, and let $\mu_i \leqslant \sigma_i$ be such that μ_1 and μ_2 are equivalent. By part (i) we have $\bar{\mu}_1 = \bar{\mu}_2$. But since $\mu_i \leqslant \sigma_i$ clearly implies $\bar{\mu}_i \leqslant \bar{\sigma}_i$, then by hypothesis $\bar{\mu}_1 \perp \bar{\mu}_2$. Thus the only conclusion is that $\bar{\mu}_1 = \bar{\mu}_2 = 0 = \mu_1 = \mu_2$, and this proves σ_1 and σ_2 are disjoint.

(ii) Assume σ_1 is multiplicity-free and $\bar{\sigma}_1 \leqslant \bar{\sigma}_2$. First, we claim that every nonzero subrepresentation μ of σ_1 contains a nonzero subrepresentation equivalent to a subrepresentation of σ_2. Indeed, regarding μ as a subrepresentation of π, we have $\bar{\mu} \leqslant \bar{\sigma}_1 \leqslant \bar{\sigma}_2$ and in particular $\bar{\mu}$ is not orthogonal to $\bar{\sigma}_2$. Utilizing property (iv) already proved, the claim follows.

Observe next that, by an exhaustion argument, we may express $\sigma_1 = \sum_j \mu_j$ as an orthogonal sum of nonzero subrepresentations μ_j, each equivalent to some subrepresentation μ_j' of σ_2. (In more detail, Zorn's lemma provides a maximal family $\{\mu_j\}$ of nonzero orthogonal subrepresentations of σ_1, each of which is equivalent to a subrepresentation of σ_2, and maximality plus the preceding paragraph show that $\sum \mu_j$ must be all of σ_1.)

We claim now that $\mu_j' \perp \mu_k'$ if $j \neq k$, which is the crucial step in the argument. Indeed, μ_j and μ_k are orthogonal subrepresentations of the *multiplicity-free* representation σ_1, so in fact we have $\mu_j \circ \mu_k$. Therefore $\mu_j' \circ \mu_k'$ and by part (iv) again we conclude $\bar{\mu}_j' \perp \bar{\mu}_k'$, and in particular $\mu_j' \perp \mu_k'$. Thus we may form the subrepresentation $\sum_j \mu_j'$ of σ_2, which is clearly equivalent to $\sum_j \mu_j = \sigma_1$, and (ii) follows.

(iii) Suppose both σ_1 and σ_2 are multiplicty-free, and $\bar{\sigma}_1 = \bar{\sigma}_2$. By part (ii), σ_1 is equivalent to a subrepresentation μ of σ_2. If $\mu \neq \sigma_2$, then by considering the orthogonal complement of the range space of μ in that of σ_2 we obtain a nonzero subrepresentation $v \leqslant \sigma_2$, $v \perp \mu$. Since σ_2 is multiplicity free we have in fact $v \,\delta\, \mu$, and thus $v \,\delta\, \sigma_1$, since μ is equivalent to σ_1. By (iv), $\bar{v} \perp \bar{\sigma}_1$, and since $\bar{\sigma}_1 = \bar{\sigma}_2$ we have $\bar{v} \perp \bar{\sigma}_2$. Since $\bar{v} \geqslant \bar{\sigma}_2$ (because $v \geqslant \sigma_2$) we conclude $\bar{v} = 0$, hence $v = 0$, a contradiction. □

In general, as one would expect, the process of forming direct sums tends to increase multiplicity, while the multiplicity of subrepresentations tends to get smaller. Nevertheless the next result asserts that, under certain conditions, multiplicity is preserved in passing to subrepresentations or to direct sums.

Proposition 2.1.6

(i) *Let π be a representation of A having multiplicity n. Then every nonzero central subrepresentation of π has multiplicity n.*

(ii) *Let $\{\sigma_j : j \in J\}$ be a mutually disjoint family of subrepresentations of π such that each σ_j has multiplicity n. Then $\sum_j \sigma_j$ has multiplicity n.*

PROOF

(i) We may as well assume that π has the form $n \cdot \mu$, where μ is a multiplicity-free representation on a Hilbert space $\mathscr{H}$. Therefore, each operator $\pi(x)$ can be represented as an $n \times n$ diagonal operator matrix of the form

$$\pi(x) = \begin{pmatrix} \mu(x) & 0 & \cdots \\ 0 & \mu(x) & \\ \cdot & & \cdot \\ \cdot & & \cdot \\ \cdot & & \cdot \end{pmatrix}$$

A typical operator in the weak closure of $\pi(A)$ has a similar form, with $\mu(x)$ replaced with an element T in the weak closure of $\mu(A)$. Thus it is clear that a central projection C in $\pi(A)''$ has a matrix representation

$$C = \begin{pmatrix} E & 0 & \cdots \\ 0 & E & \\ \cdot & & \cdot \\ \cdot & & \cdot \\ \cdot & & \cdot \end{pmatrix}$$

where E belongs to $\mu(A)'$, the center of $\mu(A)''$. So if π_C (resp. μ_E) denotes the subrepresentation of π (resp. μ) defined by C (resp. E), then the preceding shows that $\pi_C = n \cdot \mu_E$. The conclusion follows because μ_E is multiplicity-free.

(ii) Assume first that $n = 1$, i.e., each σ_j is multiplicity-free, and let μ be the subrepresentation $\sum_j \sigma_j$. Let $k = \text{card } J$. Then in an obvious sense every operator $\mu(x)$, $x \in A$, has a representation as a $k \times k$ operator matrix (T_{ij}), where $T_{ij} = 0$ if $i \neq j$ and $T_{jj} = \sigma_j(x)$. A routine computation now shows that every operator in the commutant $\mu(A)'$ has a representation (S_{ij}), where for each i, j, $S_{ij}\sigma_j(x) = \sigma_i(x)S_{ij}$ for every $x \in A$. Because $\sigma_i \,\delta\, \sigma_j$ if $i \neq j$, we see from 2.1.4 that $S_{ij} = 0$ if $i \neq j$; moreover, each "diagonal" term S_{jj} must belong to the commutative von Neumann algebra $\sigma_j(A)'$. This makes it apparent that $\mu(A)'$ is abelian, as required.

In general, assume each σ_j has multiplicty $n > 1$. Then for each j we may write $\sigma_j = \sum_k \sigma_{jk}$, where for each j $\{\sigma_{jk} : k \in K\}$ is an orthogonal family of equivalent multiplicity-free subrepresentations of σ_j, and card $K = n$. For each $k \in K$ define $\mu_k = \sum_j \sigma_{jk}$. By the preceding paragraph (and the obvious fact that subrepresentations of disjoint representations are disjoint) we conclude that each μ_k is multiplicity-free, and clearly

$$\sum_k \mu_k = \sum_k \sum_j \sigma_{jk} = \sum_j \sigma_j.$$

Since the μ's are all equivalent, $\sum_j \sigma_j$ has multiplicity card $K = n$. □

We now define the class of representations to which multiplicity theory applies.

Definition 2.1.7. A representation π is called type I if every nonzero central subrepresentation of π contains a nonzero multiplicity-free subrepresentation.

Thus, π is type I in case every central subrepresentation of π contains an irreducible subrepresentation (this is always the case when π acts on a finite-dimensional space); and if π is a factor representation (i.e., $\pi(A)'$ has trivial center) then it is type I if it contains an irreducible subrepresentation. In the event A is abelian, then *every* representation of A is type I. For every representation of A clearly contains a cyclic subrepresentation (a representation π of A on $\mathscr{H}$ is cyclic if there is a vector $\xi \in \mathscr{H}$ for which $[\pi(A)\xi] = \mathscr{H}$), and as we will see in the next section (corollary of 2.2.4), a cyclic representation of an abelian C^*-algebra is mutliplicity-free. More generally, we will see later on that every GCR algebra has only type I representations (cf. Theorem 2.4.1).

We now come to a main result.

Theorem 2.1.8 Decomposition theorem. *Let π be a type I representation of a C^*-algebra on a separable Hilbert space, and let N be the set of all positive cardinals not exceeding $\aleph_0$. Then there is an orthogonal family $\{\pi_n : n \in N\}$ of central subrepresentations of π such that*

(i) *π_n is either 0 or has multiplicity n;*
(ii) $\pi = \sum_n \pi_n$.

The family $\{\pi_n\}$ is uniquely determined by these conditions. Moreover, for each $n \in N$, the central subrepresentations of π having multiplicity n are precisely the central subrepresentations of π_n.

Proof. Fix $n \in N$. If there is no central subrepresentation of π having multiplicity n, set $\pi_n = 0$. Otherwise, by Zorn, we may choose a maximal family $\{\mu_i\}$ of nonzero orthogonal central subrepresentations of π such that each μ_i has multiplicity n. Put $\pi_n = \sum_i \mu_i$. π_n is clearly a central subrepresentation, and by 2.1.5 (iv) and 2.1.6 (ii) π_n has multiplicity n.

Now if σ is a central subrepresentation of π_n then 2.1.6 (i) shows that σ has multiplicity n. On the other hand, if σ is a central subrepresentation of π which has multiplicity n, then we claim that $\sigma \leqslant \pi_n$. For if the range space of σ is not contained in that of π_n, then since their respective projections commute, the range space of σ has nontrivial intersection with the orthogonal complement of the range space of π_n. That intersection determines a nonzero central subrepresentation of π, orthogonal to $\pi_n = \sum_i \mu_i$, which is also a subrepresentation of σ and therefore by 2.1.6 (i) has multiplicity n. But this contradicts the maximality of $\{\mu_i\}$. Thus $\sigma \leqslant \pi_n$ as asserted. This proves the last sentence of the theorem, and a moment's thought shows how the uniqueness statement follows from it.

It remains to prove that $\sum_n \pi_n = \pi$. Assume not, and let π_0 be the central subrepresentation of π defined by the orthogonal complement of the range space of $\sum_n \pi_n$. Because π is type I, π_0 contains a nonzero multiplicity-free subrepresentation σ. Utilizing Zorn's lemma once again, choose a maximal family $\{\sigma_j : j \in J\}$ of mutually orthogonal subrepresentations of π_0, each equivalent to σ. Note that J is countable, by the separability hypothesis. We will produce a nonzero subrepresentation λ of $\sum \sigma_j$ which is central (in π and therefore in $\sum \sigma_j$). This will complete the proof, for by 2.1.6 (i) λ has multiplicity $n = \text{card } J$ and so $\lambda \leqslant \pi_n$ by the preceding paragraph, contradicting the fact that by construction λ is orthogonal to every π_n. To define λ, let μ be the subrepresentation of π defined by the orthogonal complement of the range space of $\sum \sigma_j$ in π_0, and let λ be the central subrepresentation defined by the orthogonal complement of the range space of $\bar{\mu}$ in π_0. Clearly $\lambda \leqslant \sum \sigma_j$. If $\lambda = 0$, then $\bar{\mu} = \pi_0$ and hence $\bar{\sigma} \leqslant \bar{\mu}$. By 2.1.5 (ii) σ is equivalent to a subrepresentation σ' of μ, but then σ' contradicts maximality of $\{\sigma_j\}$. Thus, $\lambda \neq 0$, and the proof is complete. □

We are now in a position to give the promised reduction from type I representations to multiplicity-free representations. First, note that a representation having multiplicity n uniquely determines its multiplicity free summand to equivalence. That is, let $\pi = \sum_i \mu_i = \sum_j \nu_j$, where the $\{\mu_i\}$ (resp. $\{\nu_j\}$) are mutually orthogonal, mutually equivalent multiplicity-free subrepresentations. Then we claim: each μ_i is equivalent to each ν_j. For by 2.1.5 (i) and the fact that the μ's are all equivalent we have $\bar{\mu}_i = \bar{\mu}_k$, and in particular $\mu_i \leqslant \bar{\mu}_k$ for all i, k. Hence $\pi = \sum \mu_i \leqslant \bar{\mu}_k$ so that $\pi = \bar{\mu}_k$ for all k. Similarly $\pi = \bar{\nu}_j$ for all j. By 2.1.5 (iii) it follows that μ_k and ν_j are equivalent.

The most general type I represention (on a separable space) can be constructed as follows. Let N be the set of all cardinal numbers n, $1 \leqslant n \leqslant \aleph_0$, and let $S = \{\sigma_n : n \in N\}$ be a sequence of separable representations of A having the following properties:

(i) *each* σ_n *is* 0 *or multiplicity-free*;
(ii) $\sigma_m \mathbin{\circ\!\!\!/} \sigma_n$ *if* $m \neq n$.

Of course, the Hilbert space $\mathscr{H}_n$ of σ_n may vary with n. For each n we form the separable space $n \cdot \mathscr{H}_n = \mathscr{H}_n \oplus \mathscr{H}_n \oplus \cdots$, the direct sum of n copies of $\mathscr{H}_n$, and we consider the representation $n \cdot \sigma_n$ on $n \cdot \mathscr{H}_n$. Note that $n \cdot \sigma_n$ has multiplicity n or is the 0 representation. Finally, define a representation π_S on $1 \cdot \mathscr{H}_1 \oplus 2 \cdot \mathscr{H}_2 \oplus 3 \cdot \mathscr{H}_3 \oplus \cdots \oplus \aleph_0 \cdot \mathscr{H}_{\aleph_0}$ by

$$\pi_S(x) = 1 \cdot \sigma_1(x) \oplus 2 \cdot \sigma_2(x) \oplus \cdots \oplus \aleph_0 \cdot \sigma_{\aleph_0}(x),$$

$x \in A$. Because of the disjointness condition (ii) it follows (2.1.5) that the subrepresentations $\pi_n = n \cdot \sigma_n$ of π_S form the canonical central decomposition of π_S described in Theorem 2.1.8.

Now let $S' = \{\sigma_n' : n \in N\}$ be another family satisfying (i) and (ii). Let us say S and S' are *equivalent* if σ_n and σ_n' are equivalent for each n. It is clear that if S and S' are equivalent then so are the representations π_S and $\pi_{S'}$. Conversely, suppose π_S and $\pi_{S'}$ are equivalent. By replacing S' with an equivalent family we can assume $\pi_{S'} = \pi_S$. By the uniqueness statement in 2.1.8 we have $n \cdot \sigma_n = n \cdot \sigma_n'$ for all n, and by the discussion preceding the last paragraph this entails the equivalence of σ_n and σ_n'. It follows that S and S' are equivalent. Finally, it is an obvious consequence of 2.1.8 that *every* type I representation of A has the form π_S, for some sequence S satisfying (i) and (ii). To summarize, we have proved the following

Corollary 2.1.9. *$S \leftrightarrow \pi_S$ defines a bijective correspondence between equivalence classes of sequences S satisfying* (i) *and* (ii) *and the equivalence classes of type I representations of A on separable Hilbert spaces.*

Thus the classification problem for type I separable representations is reduced to the problem of classifying multiplicity-free representations. We remark that with only minor variations the entire analysis of this section may be carried out for representations on Hilbert spaces which are inseparable.

2.2. Commutative C^*-Algebras and Normal Operators

Let N be a normal operator on a finite dimensional Hilbert space $\mathscr{H}$. A familiar theorem from linear algebra asserts that there is an orthonormal base for $\mathscr{H}$ with respect to which the matrix of N is diagonal, that is

$$N = \sum_{j=1}^{n} \lambda_j P_j,$$

where $P_1, \ldots, P_n$ are mutually orthogonal one-dimensional projections and $\lambda_1, \ldots, \lambda_n$ range over $\operatorname{sp}(N)$ (perhaps with repetitions). Observe that the projections P_j occuring in this representation are not unique (for example, consider the identity operator as N). However, if we collect the P_j's corresponding to a single value of $\lambda \in \operatorname{sp}(N)$, then the expression becomes

$$N = \sum_{\lambda \in \operatorname{sp}(A)} \lambda E_\lambda,$$

where $\{E_\lambda : \lambda \in \text{sp}(N)\}$ is now merely a mutually orthogonal family of *nonzero* projections having sum 1. In this form the expression on the right is uniquely determined by N (note that the E_λ's are the minimal spectral projections of N), and we may define the *multiplicity* of $\lambda \in \text{sp}(N)$ by $m_N(\lambda) = \dim E_\lambda$. The function $m_N : \text{sp}(N) \to \{1, 2, \ldots, \dim \mathscr{H}\}$ is called the *multiplicity function* of N.

Now let M be another normal operator on some finite dimensional space, say

$$M = \sum_{\lambda \in \text{sp}(M)} \lambda F_\lambda.$$

Then it is not hard to see that N and M are unitarily equivalent if, and only if, $\text{sp}(N) = \text{sp}(M)$ and $m_N(\lambda) = m_M(\lambda)$ for all λ. Thus, the finite set $\text{sp}(N)$ together with the multiplicity function m_N form a complete set of unitary invariants for finite dimensional normal operators N.

Now let N be a normal operator on an infinite dimensional Hilbert space. Corresponding to the above expression we have the spectral theorem, which asserts that N has a decomposition

$$N = \int_{\text{sp}(N)} \lambda E(d\lambda),$$

where $E(\cdot)$ is a projection-valued Borel measure on $\text{sp}(N)$. Notice however, that it is no longer clear how to define the "multiplicity" of a point $\lambda \in \text{sp}(N)$ in the very frequent case where the range of $E(\cdot)$ contains no minimal nonzero elements; indeed this is precisely the point at which the technical difficulty appears. Nevertheless, we will see that it is possible to define a function analogous to m_N above, which is uniquely determined almost everywhere $(E(d\lambda))$, such that this function, together with the set $\text{sp}(N)$ and the measure class determined by $E(\cdot)$, form a complete set of unitary invariants for N.

This analysis is carried out in the somewhat more general context of representations of commutative C^*-algebras; thus, rather than analyze the normal operator N, we analyze the identity representation of $C^*(N)$.

Let A be a separable commutative C^*-algebra, and let X be the spectrum of A. Then X is a locally compact Hausdorff space and the Gelfand map identifies A isometrically and $*$-isomorphically with $C(X)$, the C^*-algebra of all continuous complex-valued functions on X vanishing at ∞ (1.1.1). By separability of $C(X)$, X is second countable, and a theorem of Urysohn [18] implies that, in fact, X is metrizable as a separable complete metric space.

We begin by describing a construction giving multiplicity-free representations of $C(X)$. Let μ be any finite Borel measure on X. Since $C(X)$ is separable, it follows easily that the Hilbert space $L^2(X, \mu)$ is separable. Let $\mathscr{B}(X)$ denote the set of all bounded complex valued Borel functions defined on X. $\mathscr{B}(X)$ becomes a commutative C^*-algebra with the pointwise operations and the norm $\|f\| = \sup_x |f(x)|$, containing $C(X)$ as a C^*-subalgebra. Each function $f \in \mathscr{B}(X)$ gives rise to a multiplication operator L_f on $L^2(X, \mu)$, defined by $(L_f\xi)(x) = f(x)\xi(x)$, and the map $f \to L_f$ is a *representation* of $\mathscr{B}(X)$ on

$L^2(X, \mu)$. Let us write π_μ for its restriction to $C(X)$. We will see that π_μ is multiplicity-free, and we will study the way in which π_μ depends on μ.

Let $\mathscr{L}$ denote the set of *all* multiplications L_f, $f \in \mathscr{B}(X)$. $\mathscr{L}$ is clearly a commutative $\star$-algebra of normal operators.

Theorem 2.2.1. *$\mathscr{L}$, $\mathscr{L}'$, and the strong closure of $\pi_\mu(C(X))$, are identical.*

Proof. It is clear that $\pi_\mu(C(X)) \subseteq \mathscr{L} \subseteq \mathscr{L}'$; we will prove that $\mathscr{L}' \subseteq \mathscr{L}$, and $\mathscr{L}$ is contained in the strong closure of $\pi_\mu(C(X))$. Choose $T \in \mathscr{L}'$, and let 1 be the obvious constant function, regarded as an element of $L^2(X, \mu)$. Then $T1 \in L^2(X, \mu)$, and we may choose a representative g in the equivalence class of functions $T1$. We claim: g is essentially bounded and $T = L_g$. Indeed if $\xi, \eta \in L^2(X, \mu)$ and ξ is bounded, then $|\int g(x)\xi(x)\bar{\eta}(x)\, d\mu(x)| = |(L_\xi T1, \eta)| = |(TL_\xi 1, \eta)| = |(T\xi, \eta)| \leqslant \|T\| \cdot \|\xi\|_2 \cdot \|\eta\|_2$. This inequality implies that $|g(x)| \leqslant \|T\|$ a.e. (μ) so by modifying g on a set of measure zero we can assume $|g(x)| \leqslant \|T\|$ for all $x \in X$. Thus $g \in \mathscr{B}(X)$. Since L_g and T are bounded, $L_g = T$ will follow if we show that $L_g = T$ on the dense set of all bounded functions in $L^2(X, \mu)$. But if $\xi \in \mathscr{B}(X)$, then $L_g \xi = \xi \cdot g = L_\xi g = L_\xi T_1 = TL_\xi 1 = T\xi$, as required. Thus $\mathscr{L}' \subseteq \mathscr{L}$.

To see that each multiplication L_g, $g \in \mathscr{B}(X)$, is in the strong closure of $\pi_\mu(C(X))$, choose a bounded sequence of functions $f_n \in C(X)$ such that $f_n(x) \to g(x)$ a.e. (μ). The dominated convergence theorem implies that $f_n(x)\xi(x) \to g(x)\xi(x)$ in L^2, for every $\xi \in L^2$, and hence $L_{f_n} \to L_g$ strongly.

As an immediate consequence we see that the commutant of $\pi_\mu(C(X))$ is abelian (namely $\mathscr{L}'' = \mathscr{L}$), and therefore *$\pi_\mu$ is multiplicity-free.*

Recall that two finite Borel measures μ, ν are *singular* (written $\mu \perp \nu$) if there is a Borel set $E \subseteq X$ so that $\mu(E) = \nu(X \backslash E) = 0$, i.e., μ is concentrated on $X \backslash E$ and ν is concentrated on E. $\mu \ll \nu$ means as usual that $\nu(E) = 0$ implies $\mu(E) = 0$ for every Borel set $E \subseteq X$, and finally μ and ν are *equivalent* if they have the same null sets (i.e., $\mu \ll \nu$ and $\nu \ll \mu$). It is evident that for two singular measures μ and ν, the only possible measure σ satisfying $\sigma \ll \mu$ and $\sigma \ll \nu$ is $\sigma = 0$. Conversely, by applying the Legesgue decomposition in a straightforward manner, the reader can easily see that if μ and ν are *not* mutually singular then there is a nonzero measure σ satisfying $\sigma \ll \mu$ and $\sigma \ll \nu$. We come now to an elegant characterization of unitary equivalence and disjointness in the class of representations $\{\pi_\mu\}$.

Theorem 2.2.2. *Let μ and ν be two finite Borel measures on X. Then π_μ is equivalent to π_ν iff μ and ν are equivalent measures. π_μ is disjoint from π_ν iff $\mu \perp \nu$.*

Proof. Assume first that μ and ν are equivalent. We will find a unitary operator $U: L^2(X, \mu) \to L^2(X, \nu)$ such that $U\pi_\mu = \pi_\nu U$. By the Radon–Nikodym theorem and the fact that $\mu \ll \nu$, there is a nonnegative function $h \in L^1(\nu)$ such that $d\mu = h\, d\nu$. Moreover, $\nu \ll \mu$ implies that $h > 0$ a.e. (ν).

For $\xi \in L^2(X, \mu)$, define $(U\xi)(x) = (h(x))^{1/2}\xi(x)$. Note that U maps $L^2(X, \mu)$ isometrically into $L^2(X, \nu)$; indeed for $\xi \in L^2(X, \mu)$ we have $\int |U\xi|^2 \, d\nu = \int |\xi|^2 h \, d\nu = \int |\xi|^2 \, d\mu = \|\xi\|^2$. The property $U\pi_\mu = \pi_\nu U$ is evident from the definition of U, and the easiest way to see that U is onto is to note by a symmetric argument that the map $V: \eta \to h^{-1/2}\eta$ is an isometry of $L^2(X, \nu)$ into $L^2(X, \mu)$ which is both a right and left inverse of U. Thus π_μ and π_ν are equivalent.

Now assume $\mu \perp \nu$. We will prove that $\pi_\mu \circ \pi_\nu$. First, choose a Borel set E such that $\mu(X \backslash E) = \nu(E) = 0$. Let f be the characteristic function of E, and consider the measure $\mu + \nu$. Then 4.1.1 shows that $L_f \in \pi_{\mu+\nu}(A)''$. Now since μ is concentrated on E and ν is concentrated on $X \backslash E$, we may regard $L^2(X, \mu)$ and $L^2(X, \nu)$ as $L^2(E, \mu)$ and $L^2(X \backslash E, \nu)$ respectively. Thus $L^2(X, \mu + \nu)$ can be identified with $L^2(E, \mu) \oplus L^2(X \backslash E, \nu) = L^2(X, \mu) \oplus L^2(X, \nu)$, and this process identifies $\pi_{\mu+\nu}$ with $\pi_\mu \oplus \pi_\nu$ in such a way that L_f becomes the projection $I \oplus 0$ of $L^2(X, \mu) \oplus L^2(X, \nu)$ onto its first coordinate space. In short, we have shown that $I \oplus 0$ belongs to $(\pi_\mu \oplus \pi_\nu)(A)''$. By 2.1.4 (iii) we conclude that $\pi_\mu \circ \pi_\nu$.

Conversely, suppose $\pi_\mu \circ \pi_\nu$. Now if μ and ν are not singular, then there is a nonzero measure σ such that $\sigma \ll \mu$ and $\sigma \ll \nu$. We will show that π_σ is equivalent to a subrepresentation of π_μ (and, by symmetry, π_σ is also equivalent to a subrepresentation of π_ν). This will clearly contradict $\pi_\mu \circ \pi_\nu$. By the Radon–Nikodym theorem there is a nonnegative Borel function $h \in L^1(X, \mu)$ such that $d\sigma = h \, d\mu$. Let E be the Borel set $\{x \in X : h(x) > 0\}$. Now $L^2(E, \mu)$ is a closed subspace of $L^2(X, \mu)$ invariant under $\pi_\mu(A)$, and thus defines a subrepresentation of π_μ. We claim that this subrepresentation is equivalent to π_σ. Indeed for each function $\xi \in L^2(X, \sigma)$, define a function $U\xi$ on E by $(U\xi)(x) = h(x)^{1/2}\xi(x)$. Then we have

$$\int_E |U\xi(x)|^2 \, d\mu(x) = \int_X |\xi(x)|^2 h(x) \, d\mu(x) = \int_X |\xi(x)|^2 \, d\sigma(x),$$

and it follows that U is a unitary map of $L^2(X, \sigma)$ onto $L^2(E, \mu)$. Evidently $U\pi_\sigma(f) = \pi_\mu(f)U$ for all $f \in A$, and the claim is proved.

It remains to prove that $\pi_\mu \sim \pi_\nu$ implies μ and ν are equivalent measures. Contrapositively, let us assume for definiteness that μ is not absolutely continuous with respect to ν. Then by the Lebesgue decomposition we can write $\mu = \mu_0 + \sigma$, where $\sigma \perp \nu$, $\sigma \neq 0$, and $\mu_0 \ll \nu$. Of course $\sigma \ll \mu$. Now by the preceding paragraph π_σ is a nonzero representation equivalent to a subrepresentation of π_μ. On the other hand $\sigma \perp \nu$ implies $\pi_\sigma \circ \pi_\nu$, by an argument we have already gone through. Thus π_μ contains a nonzero subrepresentation *disjoint* from π_ν, and thus π_μ and π_ν could not be equivalent. □

Thus the unitary equivalence class of a particular π_μ depends not on the particular measure μ but on the *equivalence class* of all finite measures which are equivalent to μ. Let us call such an equivalence class of measures a *measure class*. As a simple example, let μ be an *atomic* measure; i.e., there is a countable (or finite) subset $\{x_1, x_2, \ldots\}$ of X such that $\mu(\{x_i\}) > 0$ for all i, and $\mu(X\backslash\{x_1, x_2, \ldots\}) = 0$. In this case a measure v belongs to the measure class determined by μ iff v shares these two properties with μ. Thus the measure classes determined by atomic measures correspond 1—1 with the countable subsets of X. Two such measures classes are mutually singular iff their corresponding countable subsets are disjoint. This explains why, even in the nonatomic case, it is convenient to think of measure classes as "generalized" subsets of X. Indeed the reader may wish to discover for himself the precise sense in which the measure classes can be made into a Boolean σ-ring (without identity).

We will now show that the above family $\{\pi_\mu\}$ of multiplicity-free representations exhausts all of the (separable) possibilities. First, a general lemma.

Lemma 2.2.3. *Let A be a C^*-algebra and let π be a nondegenerate multiplicity-free representation of A on a separable Hilbert space $\mathscr{H}$. Then there is a vector $\xi \in \mathscr{H}$ such that $[\pi(A)\xi] = \mathscr{H}$.*

Note that every nonzero $\pi(A)$-invariant subspace $\mathscr{M}$ contains a nonzero vector ξ such that $[\pi(A)\xi] \subseteq \mathscr{M}$ (any $\xi \neq 0$ in $\mathscr{M}$ will do). By Zorn's lemma, then, we may find a set $\{\xi_i : i \in I\}$ of unit vectors in $\mathscr{H}$ which is maximal with respect to the property $\pi(A)\xi_i \perp \pi(A)\xi_j$; for $i \neq j$ in I. Note that $\sum_i [\pi(A)\xi_i] = \mathscr{H}$ (for otherwise any unit vector in $(\sum_i [\pi(A)\xi_i])^\perp$ would contradict maximality of $\{\xi_i\}$). Since $\mathscr{H}$ is separable, I is countable, and we may enumerate the $\{\xi_i\}$ as a sequence $\xi_1, \xi_2, \ldots$. Let $\xi = \sum_n 2^{-n}\xi_n$. To see that ξ is cyclic, it suffices to show that each ξ_n belongs to $[\pi(A)\xi]$ (because $\mathscr{H} = \sum_n [\pi(A)\xi_n]$). But the projection P_n on $[\pi(A)\xi_n]$ belongs to $\pi(A)'$, and since $\pi(A)' \subseteq \pi(A)''$ because $\pi(A)'$ is abelian, we conclude from the double commutant theorem that P_n belongs to the strong closure of $\pi(A)$. Thus $P_n\xi \in [\pi(A)\xi]$, and since $P_n\xi = 2^{-n}\xi_n$, the proof is complete. □

We remark that the converse of this lemma is valid for *commutative* C^*-algebras, but false in general. See the corollary following 2.2.4.

Theorem 2.2.4. *Let A be a commutative separable C^*-algebra and let π be a nondegenerate multiplicity-free representation of A on a separable Hilbert space. Then there is a finite Borel measure μ on X, the spectrum of A, such that π is equivalent to π_μ.*

Proof. We may identify A with $C(X)$ via the Gelfand map. Let $\mathscr{H}$ be the representation space of π. Then by 2.2.3 there is a nonzero vector $\xi \in \mathscr{H}$ such that $\mathscr{H} = [\pi(A)\xi]$. Define a bounded linear functional ρ on $C(X)$ by $\rho(f) = (\pi(f)\xi, \xi)$. ρ is clearly positive [i.e., $\rho(|f|^2) = (\pi(f)^*\pi(f)\xi, \xi) = \|\pi(f)\xi\|^2 \geqslant 0$] so by the Riesz–Markov theorem there is a positive Borel measure μ on X such that $\rho(f) = \int f(x)\, d\mu(x)$, $f \in C(X)$. Now regarding $C(X)$ as a dense linear subset of $L^2(X, \mu)$, we define a map $U : C(X) \to \mathscr{H}$ by $Uf = \pi(f)\xi$. Note that $\|Uf\|^2 = \|\pi(f)\xi\|^2 = (\pi(f)\xi, \pi(f)\xi) = (\pi(|f|^2)\xi, \xi) = \rho(|f|^2) = \int |f|^2\, d\mu$, so that U is an isometry. Since its domain and range are dense it extends to a unitary map of $L^2(X, \mu)$ onto $\mathscr{H}$. To see that $U\pi_\mu = \pi U$, choose $f, g \in C(X)$. Then regarding g as an element of $L^2(X, \mu)$ we see that $U\pi_\mu(f)g = U(f \cdot g) = \pi(fg)\xi = \pi(f)\pi(g)\xi = \pi(f)Ug$. The required formula follows because $C(X)$ is dense in $L^2(X, \mu)$. □

We remark that the above results used separability of A in only a superficial way, and in fact the results are valid, in only slightly modified form, for separable multiplicity-free representations of arbitrary commutative C^*-algebras. In place of Borel measures on X, for example, one deals with finite regular Borel measures μ which are "separable" in the sense that $L^2(X, \mu)$ is separable. One may also extend the results to representations on inseparable Hilbert spaces, at the expense of finiteness of the measure μ. For the somewhat fussy details, see ([7], Chapter I, Section 7).

Corollary. *Let $\mathscr{H}$ be a separable Hilbert space and let π be a representation of a separable abelian C^*-algebra on $\mathscr{H}$. Then π is multiplicity-free if and only if π has a cyclic vector.*

An abelian von Neumann algebra $\mathscr{R}$ on $\mathscr{H}$ is maximal abelian (i.e., $\mathscr{R}' = \mathscr{R}$) if and only if $\mathscr{R}$ has a cyclic vector.

Proof. Note that the second paragraph follows from the first. For by Proposition 1.2.3, every abelian von Neumann algebra $\mathscr{R}$ on $\mathscr{H}$ is the strong closure of an abelian separable C^*-subalgebra $\mathscr{A}$; and in this event $\mathscr{R}' = \mathscr{R}$ iff $\mathscr{R}' = \mathscr{A}'$ is commutative iff the identity representation of $\mathscr{A}$ is multiplicity-free.

For the first paragraph, the "only if" part is Lemma 2.2.3, and the converse follows from 2.2.4 and the remark following 2.2.1. □

Let us now summarize these classification results, and apply them to obtain a complete set of unitary invariants for normal operators on separable Hilbert spaces.

Let A be a separable commutative C^*-algebra, and let X be the spectrum of A. Identify A with $C(X)$. Let us construct a representation of $C(X)$ as follows. Choose a sequence $\mu_\infty, \mu_1, \mu_2, \ldots$ of finite Borel measures on X such that $\mu_i \perp \mu_j$ if $i \neq j$ (some, but not all, of the μ_i are allowed to be 0). Owing to 2.2.1 and 2.2.2, $\pi_{\mu_\infty}, \pi_{\mu_1}, \pi_{\mu_2}, \ldots$ forms a sequence of multiplicity-free disjoint separable representations of $C(X)$, and thus $\pi = \infty \cdot \pi_{\mu_\infty} \oplus 1 \cdot \pi_{\mu_1} \oplus 2 \cdot \pi_{\mu_2} \oplus \cdots$ defines a separable representation of $C(X)$ which is exhibited in its unique canonical decomposition as in 2.1.9. Moreover, by 2.1.9 and the preceding results of this section, *every separable representation of $C(X)$ has this form.* Finally, given two sequences $\mu_\infty, \mu_1, \ldots, \nu_\infty, \nu_1, \ldots$ of finite measures as above, then the associated representations are equivalent if, and only if, μ_n and ν_n are equivalent measures, for every $n = \infty, 1, 2, \ldots$ (this, by 2.1.9 and 2.2.2). Thus, the separable representations of $C(X)$ are completely classified to unitary equivalence by sequences $C_\infty, C_1, C_2, \ldots$ of (finite) measure classes on X satisfying $C_i \perp C_j$ if $i \neq j$.

Now fix $\mu_\infty, \mu_1, \mu_2, \ldots$ as above. We want to sketch an alternate description of the representation $\pi = \infty \cdot \pi_{\mu_\infty} \oplus 1 \cdot \pi_{\mu_1} \oplus \cdots$ which is somewhat more revealing. There is clearly no essential change if we renormalize each μ_j so that $\sum_j \mu_j(X) < \infty$, and in this case $\mu(E) = \mu_\infty(E) + \mu_1(E) + \mu_2(E) + \cdots$ defines a finite Borel measure on X. Now since $\mu_i \perp \mu_j$ if $i \neq j$, we can find a disjoint sequence $E_\infty, E_1, E_2, \ldots$ of Borel sets having union X, such that $\mu_i(F) = \mu(F \cap E_i)$ for every Borel set F and every $i = 1, 2, \ldots$. It follows from the singularity properties of $\{\mu_i\}$ that $\mu_i(E_j) = 0$ if $i \neq j$, and moreover the partition $\{E_i\}$ is uniquely determined by these properties modulo sets of μ-measure zero. If one thinks of E_n roughly as being the set on which the multiplicity of π equals n, then the following construction should become transparent. For each n, $1 \leqslant n \leqslant \infty$, choose an n-dimensional Hilbert space $\mathscr{K}_n$ (where of course ∞ is taken as $\aleph_0$). Let $L^2(E_n, \mu; \mathscr{K}_n)$ denote the vector space of all functions $\xi: E_n \to \mathscr{K}_n$, which are Borel functions in the sense that $x \mapsto (\xi(x), u)$ is a Borel function for each $u \in \mathscr{K}_n$, and which satisfy $\|\xi\|^2 = \int_{E_n} \|\xi(x)\|^2 \, d\mu(x) < \infty$. After identifying functions which agree almost everywhere (μ), $L^2(E_n, \mu; \mathscr{K}_n)$ becomes a Hilbert space. As one would expect, the direct sum $\mathscr{H}$ of these Hilbert spaces can be identified with the Hilbert space of all Borel functions ξ from X into the disjoint union $\mathscr{K}_\infty \cup \mathscr{K}_1 \cup \mathscr{K}_2 \cup \ldots$, which satisfy $\xi(x) \in \mathscr{K}_n$ for $x \in E_n$ and for which the norm $\|\xi\|^2 = \int_X \|\xi(x)\|^2 \, d\mu(x)$ is finite. Then for each $f \in C(X)$ we can define a multiplication operator $\pi(f)$ on $\mathscr{H}$ by $(\pi(f)\xi)(x) = f(x)\xi(x)$, $x \in X$, and thus π is a representation of $C(X)$ on $\mathscr{H}$. (For the details, and related results, see Section 4.2.) It is more or less obvious that π is equivalent to $\infty \cdot \pi_{\mu_\infty} + 1 \cdot \pi_{\mu_1} + 2 \cdot \pi_{\mu_2} + \cdots$, which is to say that the restriction of π to each subspace $L^2(E_n, \mu; \mathscr{K}_n)$ is equivalent to $n \cdot \pi_{\mu_n}$, $n = \infty, 1, 2, \ldots$.

Now for each $x \in X$, let us define the "multiplicity" $m(x)$ of π at x as follows, $m(x) = k$ if $x \in E_k$, $k = \infty, 1, 2, \ldots$. Thus $m(\cdot)$ is a Borel function from X into the set $\{\infty, 1, 2, \ldots\}$ of denumerable cardinals, for which $E_k = \{x \in X : m(x) = k\}$. Since the partition $\{E_k\}$ is unique modulo sets of μ-measure zero, the multiplicity function $m(\cdot)$ is uniquely defined to a μ-equivalence. Conversely, given a positive finite measure μ on X and a Borel function $m(\cdot): X \to \{\infty, 1, 2, \ldots\}$, we obtain a partition $\{E_\infty, E_1, E_2, \ldots\}$ of X by taking $E_k = \{x \in X : m(x) = k\}$, and the family $\{\mu_k\}$ of mutually singular measures is recovered by taking $\mu_k(S) = \mu(S \cap E_k)$, $S \subseteq X$, $k = \infty, 1, 2, \ldots$.

This description of π is conveniently expressed in terms of direct integral notation:

$$\pi = \int_X^\oplus m(x) \cdot x \, d\mu(x),$$

where the symbol $n \cdot x$ ($n = \infty, 1, 2, \ldots$, $x \in X$) suggests the direct sum of n copies of the one-dimensional representation $x: f \in C(X) \to f(x) \in \mathbb{C}$. Of course we have not given a formal definition of direct integrals; the right side of this equation suggests nothing more than the construction described above, given a pair (μ, m) consisting of a measure μ and a multiplicity function m. The results of this section can therefore be restated as follows. *Every separable representation of* $C(X)$ *is equivalent to a direct integral* $\int^{\oplus} m(x) \cdot x \, d\mu(x)$, *for some measure* μ *on* X *and some Borel multiplicity function* $m: X \to \{\infty, 1, 2, \ldots\}$. *Given two such pairs* (μ, m) *and* (v, n), *then* $\int^{\oplus} m(x) \cdot x \, d\mu(x)$ *and* $\int^{\oplus} n(x) \cdot x \, dv(x)$ *are equivalent if, and only if,* μ *and* v *are equivalent measures and* $m(x) = n(x)$ *almost everywhere.*

We will now interpret these results so as to obtain unitary invariants for normal operators. Let X be a fixed compact set in the complex plane and let us classify all normal operators N having spectrum X which act on separable Hilbert spaces. For such an N let A be the C^*-algebra generated by N and the identity. Let $\zeta \in C(X)$ be the function $\zeta(z) = z$, $z \in X$. Then the functional calculus provides us with a unique representation π_N of $C(X)$ satisfying $\pi_N(\zeta) = N$, and if M is another such normal operator then M and N are unitarily equivalent iff π_M and π_N are equivalent representations of $C(X)$. Applying the above to π_N, we obtain a finite measure v on X and a multiplicity function $n: X \to \{\infty, 1, 2, \ldots\}$ such that $\pi_N = \int_X^{\oplus} n(z) \cdot z \, dv(z)$. In particular, note that this provides a representation for N as "multiplication by ζ" in a certain L^2-space of vector-valued functions on X. If we denote this symbolically by

$$N = \int_X^{\oplus} n(z) \cdot \zeta(z) \, dv(z),$$

then we may conclude: *N is unitarily equivalent to* $M = \int_X^{\oplus} m(z) \cdot \zeta(z) \, d\mu(z)$ *if, and only if,* μ *and* v *define the same measure class* $(\mu \sim v)$ *and their multiplicity functions agree almost everywhere* $[m(x) = n(x)$ a.e. $(\mu)]$. In the special case where X is contained in the real axis, the above is equivalent to the Hahn–Hellinger classification of bounded self-adjoint operators.

2.3. An Application: Type I von Neumann Algebras

Let $\mathscr{R}$ be a von Neumann algebra acting on a separable Hilbert space $\mathscr{H}$. Assume that $\mathscr{R}$ is a *factor*, i.e., the center $\mathscr{R} \cap \mathscr{R}'$ consists only of scalars. A projection $E \in \mathscr{R}$ is called *minimal* if the only projections $F \in \mathscr{R}$ satisfying $0 \leqslant F \leqslant E$ are $F = 0$ or $F = E$. Note that since $\mathscr{R}$ is a factor this is equivalent to the requirement that every subprojection of E have the form CE for some central projection $C \in \mathscr{R} \cap \mathscr{R}'$. When $\mathscr{R}$ is not a factor, the appropriate counterparts of *minimal* projections turn out to be projections E all of whose subprojections (in $\mathscr{R}$) are "central slices" of E, that is, of the form CE with C a central projection. Now fix E and consider the family of operators $E\mathscr{R}E = \{ETE: T \in \mathscr{R}\}$. $E\mathscr{R}E$ is clearly a $*$-algebra, and it turns out to be weakly closed as well. The process of passing from $\mathscr{R}$ to $E\mathscr{R}E$ is called *localization.* Now it is not hard to see (though we do not require this fact) that the above condition on E is satisfied if, and only if, the local von Neumann algebra $E\mathscr{R}E$ is abelian. When E has the latter property it is called an *abelian* projection.

Here is a simple example of an abelian projection. Choose any abelian von Neumann algebra $\mathscr{A}$ on a Hilbert space $\mathscr{H}$ and let $\mathscr{R}$ be the von Neumann algebra of all $n \times n$ matrices over $\mathscr{A}$, $n = 1, 2, \ldots$ (if one wishes, he may cause $\mathscr{R}$ to act in the obvious way as operators on the direct sum of n copies of $\mathscr{H}$). Then for every projection $P \in \mathscr{A}$, the operator matrix E defined by

$$E = \begin{pmatrix} P & & & \\ & 0 & & 0 \\ & & 0 & \\ & & & \ddots & \\ & 0 & & & 0 \end{pmatrix}$$

is an abelian projection in $\mathscr{R}$.

Definition 2.3.1. A von Neumann algebra $\mathscr{R}$ is called type I if every nonzero central projection in $\mathscr{R}$ dominates a nonzero abelian subprojection in $\mathscr{R}$.

Thus a factor is type I iff it contains a nonzero minimal projection. At the other extreme, an abelian von Neumann algebra is always type I, for every one of its projections is abelian. We also remark that if one deletes the term "central" from the above definition the resulting definition turns out to be equivalent [7]. We will deduce the structure theory for type I von Neumann algebras from the results of Section 2.1. The proofs are only sketched, but we mention all the steps and the reader should be able to fill in the details easily.

There is a simple relation between abelian projections in $\mathscr{R}$ and multiplicity-free representations of $\mathscr{R}'$. For if E is an abelian projection in $\mathscr{R}$, then $\pi_E(T') = T'|_{E\mathscr{H}}$ defines a representation of $\mathscr{R}'$ on the range of E (note that π_E is a subrepresentation of the identity representation of $\mathscr{R}'$), and we claim that π_E is multiplicity-free. This will follow if we show that $\pi_E(\mathscr{R}')'$ is contained in $E\mathscr{R}|_{E\mathscr{H}}$, for the latter is clearly isomorphic to the abelian algebra $E\mathscr{R}E$. But if $T \in \mathscr{L}(E\mathscr{H})$ commutes with $\pi_E(\mathscr{R}') = \mathscr{R}'|_{E\mathscr{H}}$ then TE is an operator on $\mathscr{H}$ whose range is contained in $E\mathscr{H}$ and which commutes with $\mathscr{R}'$ (thus $TE \in E\mathscr{R}$), and therefore $T = TE|_{E\mathscr{H}} \in E\mathscr{R}|_{E\mathscr{H}}$, as required. Though we shall not require it, the converse is also true, so that $E\mathscr{R}|_{E\mathscr{H}}$ equals the commutant of $\mathscr{R}'|_{E\mathscr{H}}$ ([7], p. 18).

The following theorem implies that the commutant of a type I von Neumann algebra is type I.

Theorem 2.3.2. *Let $\mathscr{R}$ be a von Neumann algebra on a separable Hilbert space. Then the following are equivalent:*

(i) *The identity representation of $\mathscr{R}$ is type I.*
(ii) *$\mathscr{R}$ is a type I von Neumann algebra.*
(iii) *The identity representation of $\mathscr{R}'$ is type I.*
(iv) *$\mathscr{R}'$ is a type I von Neumann algebra.*

Proof. By symmetry and the double commutant theorem it suffices to prove (i) $\Rightarrow$ (ii) $\Rightarrow$ (iii). And since (ii) $\Rightarrow$ (iii) follows from the preceding remarks, we are left with (i) $\Rightarrow$ (ii).

Assume (i), and let C be a nonzero central projection of $\mathscr{R}$. Applying 2.1.8 to the identity representation of $\mathscr{R}$, we find orthogonal central projections $C_\infty, C_1, C_2, \dots$ having sum I such that the subrepresentation $T \in \mathscr{R} \mapsto T|_{C_n\mathscr{H}}$ has multiplicity n whenever $C_n \neq 0$. Choose an n such that $CC_n \neq 0$. By passing from C to CC_n and from $\mathscr{H}$ to $C_n\mathscr{H}$ we may assume that the identity representation of $\mathscr{R}$ has multiplicity n. Thus there is a Hilbert space $\mathscr{H}_0 \neq 0$ and a von Neumann algebra $\mathscr{R}_0$ on $\mathscr{H}_0$ such that $\mathscr{R}_0'$ is abelian and such that $\mathscr{H} = \mathscr{H}_0 \oplus \mathscr{H}_0 \oplus \cdots$ (n times) and $\mathscr{R}$ consists of all operators of the form $T_0 \oplus T_0 \oplus \cdots$ (n times) with $T_0 \in \mathscr{R}_0$. Since the map $T_0 \in \mathscr{R}_0 \mapsto T_0 \oplus T_0 \oplus \cdots \in \mathscr{R}$ is clearly a $*$-isomorphism, the problem has been reduced to finding a nonzero abelian projection in $\mathscr{R}_0$.

But that is easy. Choose any nonzero vector $\xi \in \mathscr{H}_0$, and let $E \in \mathscr{L}(\mathscr{H})$ be the projection onto $[\mathscr{R}_0'\xi]$. Because the range of E is $\mathscr{R}_0'$-invariant, E belongs to $\mathscr{R}_0'' = \mathscr{R}_0$. Moreover, since $E\mathscr{H}_0$ is a *cyclic* subspace for the abelian von Neumann algebra $\mathscr{R}_0'|_{E\mathscr{H}_0}$ it follows from the corollary of 2.2.4 that the commutant of $\mathscr{R}_0'|_{E\mathscr{H}_0}$ is abelian (and in fact equals $\mathscr{R}_0'|_{E\mathscr{H}_0}$). Thus, to prove that E is an abelian projection in $\mathscr{R}_0$ it suffices to show that $E\mathscr{R}_0|_{E\mathscr{H}}$ commutes with $\mathscr{R}_0'|_{E\mathscr{H}}$. But a moment's thought shows that is obvious, and the proof is complete. □

Now let us see how this leads to a structure theorem for type I von Neumann algebras on separable Hilbert spaces. Fix $\mathscr{R}$, a type I von Neumann algebra. By 2.3.2 above, the identity representation of $\mathscr{R}'$ is a type I representation, and thus by 2.1.8 there exist central projections $C_\infty, C_1, C_2, \dots$ in $\mathscr{R}$ such that if $\mathscr{R}_n$ (resp. $\mathscr{R}_n'$) denotes the restriction of $\mathscr{R}$ (resp. $\mathscr{R}'$) to the range of C_n, then the identity representation of $\mathscr{R}_n'$ is 0 or has multiplicity n. Now $\mathscr{R}$ is clearly the direct product

$$\prod_n \mathscr{R}_n = \{T_\infty \oplus T_1 \oplus T_2 \oplus \cdots : T_n \in \mathscr{R}_n, \sup_n \|T_n\| < \infty\},$$

and each nonzero factor $\mathscr{R}_n$ has the property that its commutant $\mathscr{R}_n'$ has uniform multiplicity n. Such a von Neumann algebra is called *homogeneous of degree n*, and we have just proved that *every type I von Neumann algebra is a direct product of homogeneous von Neumann algebras of distinct degrees.* The uniqueness assertion of 2.1.8 implies that this decomposition is unique. Let us now consider a typical nonzero homogeneous factor $\mathscr{R}_n$. As in the proof of the preceding theorem, we may represent $\mathscr{R}_n'$ as the set of all operators having the form $T_0 \oplus T_0 \oplus \cdots$ (n times) where T_0 ranges over the commutant of an *abelian* von Neumann algebra $\mathscr{Z}_n$. Therefore $\mathscr{R}_n = \mathscr{R}_n''$ is represented as the algebra of all $n \times n$ operator matrices whose entries range over the abelian von Neumann algebra $\mathscr{Z}_n$ (if $n = \infty$ then of course one considers only those matrices that represent bounded operators). Now while different decompositions of the space will lead to different abelian algebras in place of $\mathscr{L}_n$, 2.1.5 (iii) implies that they are all unitarily equivalent. Conclusion: *every homogeneous von Neumann algebra of degree n "is" the algebra of all (bounded) n × n matrices over an abelian von Neumann algebra which is unique to unitary equivalence.* Thus we have a complete description of separably acting type I von Neumann algebras. As we have already suggested, there are analogous results in the inseparable case, and all the results of this section can be generalized in that direction. For an alternate and more general treatment, along with much additional information about type I von Neumann algebras, see [7]. The results of this section are due to Kaplansky [17].

2.4. GCR Algebras Are Type I

It is customary to call a C^*-algebra A *type I* if every (nondegenerate) representation of A is a type I representation. Unfortunately· this usage is in direct conflict with the terminology of the preceding section: a type I von Neumann algebra is almost never a type I C^*-algebra. For example, $\mathscr{L}(\mathscr{H})$ is a prototypical type I von Neumann algebra in the sense of 2.3.1. While on the other hand, if $\mathscr{H}$ is infinite dimensional then $\mathscr{L}(\mathscr{H})$ is known to have non type I representations, and so does not qualify as a type I C^*-algebra in the above sense. Nevertheless, both usages of the term have become so widespread that there is now no alternative but to learn to live with them. As it happens, it is almost always obvious in context just what is intended.

In any case, if one wants to use the multiplicity theory of Section 2.1 to analyze representations, it is important to know that the underlying C^*-algebra one has is type I. We conclude this chapter by sketching the proof of a theorem of Kaplansky [15] which implies that this is true for all the C^*-algebras that were encountered in Chapter 1.

Theorem 2.4.1. *Every (nondegenerate) representation of a* GCR *algebra is type I.*

Proof. Let A be a GCR algebra. Now a nondegenerate representation π of A is type I iff the identity representation of $\pi(A)''$ is type I. According to 2.3.2 then it suffices to show that $\pi(A)''$ is a type I von Neumann algebra; and a moment's thought shows that this reduces to proving that $\pi(A)''$ contains a nonzero abelian projection. Since $\pi(A)$ is isomorphic to the quotient $A/\ker \pi$, it is GCR, and thus we are reduced to proving the following assertion: every nonzero separable GCR algebra $\mathscr{A}$ acting on a Hilbert space has the property that $\mathscr{A}''$ contains a nonzero abelian projection E.

To sketch the construction of E, note that there is a nonzero element $F \in \mathscr{A}$ such that $\sigma(F)$ is compact for every irreducible representation σ of $\mathscr{A}$ (any nonzero element in the CCR ideal of $\mathscr{A}$ will do). Actually, more is true. One can even find $F \neq 0$ in $\mathscr{A}$ such that $\sigma(F)$ has rank 0 or 1 for every irreducible representation σ (for the details, see [6], Lemma 4.4.4). Choose such an F; by replacing F with F^*F we can also assume F is positive. Consider the subalgebra $F\mathscr{A}F$ of $\mathscr{A}$. Since $\sigma(F)$ has rank 1 or 0 for every such σ, all operators of the form $\sigma(F)\sigma(X)\sigma(F)$ $(X \in \mathscr{A})$ commute with one another, and therefore $\sigma(XY - YX) = 0$ for all $X, Y \in F\mathscr{A}F$ and every σ. Since the intersection of the kernels of all irreducible representations of $\mathscr{A}$ is trivial (corollary of 1.7.2) we see that $F\mathscr{A}F$ is in fact abelian, and therefore $F\mathscr{A}''F$ is abelian since it is contained in the weak closure of $F\mathscr{A}F$.

Now if F is a projection then we are done. If it is not, then choose $\varepsilon > 0$ so that the spectrum of F meets the interval $[2\varepsilon, +\infty]$, and define a Borel function $g:[0, +\infty) \to \mathbb{R}$ as follows: $g(x) = 1/x$ if $x \geqslant \varepsilon$, $g(x) = 0$ if $0 \leqslant x < \varepsilon$. Then $g(F)$ belongs to $\mathscr{A}''$, it commutes with F, and moreover $E = Fg(F)$ is a nonzero projection such that $E\mathscr{A}E$ is abelian. □

There is a converse to this theorem of Kaplansky's, namely, if a separable C^*-algebra A has only type I representations (or more simply if the factor representations of A are all type I) then A is GCR. This converse is a deep theorem of Glimm (see [13], or Section 9 of [6]).

Borel Structures 3

In this chapter we have assembled certain nontrivial results on Borel structures which are needed in the analysis of representations of noncommutative C^*-algebras. Since no exposition of this material exists in a form appropriate for our purposes and since the results are useful in other areas of mathematics, we have arranged that this chapter can be read independently of the others.

3.1. Polish Spaces

A *Polish space* is a topological space which is homeomorphic to a separable complete metric space. In this section we collect a few properties of Polish spaces for later use. Note first that a countable direct product $\prod P_i$ of Polish spaces $P_1, P_2, \ldots$ is Polish (indeed, if d_i is a metric on P_i defining its topology, then

$$d(\{x_i\}, \{y_i\}) = \sum_{i=1}^{\infty} 2^{-i} d(x_i, y_i)/(1 + d(x_i, y_i)$$

is a metric which defines the topology on $\prod_i P_i$). Similarly, a countable direct sum of Polish spaces is Polish. A closed subspace of a Polish space is obviously Polish, but it is less clear that the same is true of open subspaces:

Lemma 3.1.1. *An open subspace of a Polish space is Polish.*

Proof. Let G be an open set in P, and let d be a metric on P. Define a function $f: G \to \mathbb{R}$ by $f(x) = 1/d(x, P \backslash G)$ [$d(x, S)$ of course denotes the distance from x to the set S]. Since f is continuous on G, $x \mapsto (x, f(x))$ is a homeomorphism of G onto the graph of f, a subspace of the Polish space $P \times \mathbb{R}$. Thus it suffices to show that the graph of f is closed in $P \times \mathbb{R}$, by the preceding remarks. But if $x_n \in G$, $x \in P$, $t \in \mathbb{R}$ are such that $x_n \to x$ and $f(x_n) \to t$, then $f(x_n)$ is bounded so that x cannot belong to $P \backslash G$. Thus $x \in G$, and by continuity $t = \lim_n f(x_n) = f(x)$. □

Recall that a subset of a topological space is called a G_δ if it is a countable intersection of open sets.

Theorem 3.1.2. *Let P be a Polish space and let $S \subseteq P$ be a G_δ. Then S is a Polish space in its relative topology.*

SKETCH OF PROOF. Let $G_i \subseteq P$ be a sequence of open sets such that $S = \bigcap G_i$. By the lemma, we can find Polish spaces P_i and homeomorphisms $f_i : P_i \to G_i$. Define Q to be the set of all sequences $(p_1, p_2, \ldots) \in \prod P_i$ such that $f_1(p_1) = f_2(p_2) = \cdots$. Q is clearly a closed subset of the Polish space $\prod P_i$, and is therefore Polish. Define a function $f : Q \to P$ as follows: $f(p_1, p_2, \ldots) = f_1(p_1)$. A moment's thought shows that f maps Q onto $\bigcap G_i = S$, and it is easy to verify that f is 1—1, continuous, and its inverse is continuous (because every f_i has each of these properties). Thus S is Polish. □

A familiar result from the theory of metric spaces asserts that every closed set is a G_δ. Thus the preceding theorem includes closed subspaces as well as open ones. But these two cases of course fall far short of exhausting the possibilities; for example the set S of irrationals forms a G_δ in the set of real numbers, and therefore S is a Polish space. Finally, it is an interesting fact that the converse of 3.1.2 is true: if a subspace S of a Polish space is Polish, then S is a G_δ ([4], p. 197).

We come now to an important technical concept. Let X be a topological space. For every $k \geqslant 1$ and every k-tuple of positive integers $n_1, \ldots, n_k$, let $A_{n_1 n_2 \cdots n_k}$ be a subset of X. The family $\{A_{n_1 n_2 \cdots n_k}\}$ is called a *sieve* for X if the following properties are satisfied:

(i) $\bigcup_{n_1=1}^{\infty} A_{n_1} = X$.
(ii) $\bigcup_{\ell=1}^{\infty} A_{n_1 \cdots n_k \ell} = A_{n_1 \cdots n_k}$, *for every* $k \geqslant 1$ *and every* $n_1, \ldots, n_k \geqslant 1$.

A sieve is called *open* if each $A_{n_1 \cdots n_k}$ is an open set. The best way to understand what sieves are is to construct one:

Proposition 3.1.3. *Let P be a Polish space and let d be a metric defining the topology of P. Then P admits an open sieve $\{A_{n_1 \cdots n_k}\}$ with the following properties*:

(iii) $\operatorname{diam}(A_{n_1 \cdots n_k}) \leqslant 1/k$, *for all* $k, n_1, \ldots, n_k$;
(iv) $\bar{A}_{n_1 \cdots n_k \ell} \subseteq A_{n_1 \cdots n_k}$, *for all* $n_1, \ldots, n_k, \ell$.
(v) *Each* $A_{n_1 \cdots n_k}$ *is nonempty.*

PROOF. Since P is separable we can find a sequence $A_1, A_2, \ldots$ of nonempty open balls of diameter $\leqslant 1$ whose union is P (for example, let $\{x_1, x_2, \ldots\}$

be a countable dense set and let A_i be the open ball of radius $\frac{1}{2}$ centered at x_i). This gives us $\{A_{n_1}\}$. Now fix n_1 and consider the subspace A_{n_1}. In a similar manner we can find nonvoid open balls $A_{n_1 1}, A_{n_1 2}, \ldots$ in A_{n_1} whose union is A_{n_1}, whose diameters are all $\leqslant \frac{1}{2}$, and whose closures are all contained in A_{n_1}. Thus we now have $\{A_{n_1 n_2}\}$, and the proof is completed by an obvious induction. □

Now let $\{A_{n_1 \cdots n_k}\}$ be an open sieve with the properties (iii), (iv), and (v) above. Note first that the elements of the sieve form a base for the topology of P. Indeed if $x \in P$, $\varepsilon > 0$, and B is the open ball about x of radius ε (the metric is of course the same as the one used in 3.1.3), then for any k we have $P = \bigcup \{A_{n_1 \cdots n_k} : n_1 \geqslant 1, \ldots, n_k \geqslant 1\}$, and hence there exist $n_1, \ldots, n_k$ such that $x \in A_{n_1 \cdots n_k}$; since $\operatorname{diam}(A_{n_1 \cdots n_k}) \leqslant 1/k$ we also have $A_{n_1 \cdots n_k} \subseteq B$ so long as we choose k larger than $1/\varepsilon$, proving the assertion. Second, note that for every sequence $n_1, n_2, \ldots$ of integers, the sequence of nonvoid closed sets $\bar{A}_{n_1}, \bar{A}_{n_1 n_2}, \bar{A}_{n_1 n_2 n_3}, \ldots$ is decreasing, and by (iii) their diameters shrink to zero. So by the Cantor nested sets theorem for complete metric spaces, they intersect in a unique point. By (iv) this intersection is $\bigcap_{k=1}^{\infty} A_{n_1 \cdots n_k}$, and thus the latter intersection contains one and only one point.

Example. Let $N = \{1, 2, \ldots\}$ be the set of positive integers, endowed with the discrete topology. Then N is a Polish space (consider the metric $\delta(m, n) = 0$ or 1 according as $m = n$ or $m \neq n$), and thus the countable Cartesian product $N^\infty = N \times N \times \cdots$ is also a Polish space. Here is one good metric on N^∞: for two sequences $\{m_i\}, \{n_i\} \in N^\infty$ let $d(\{m_i\}, \{n_i\}) = \sum_{k=1}^{\infty} 2^{-k} \delta(m_k, n_k)$. For each k-tuple of integers $n_1, \ldots, n_k$ define $U_{n_1 \ldots n_k} = \{\{m_i\} \in N^\infty : m_1 = n_1, \ldots, m_k = n_k\}$. Then $\{U_{n_1 \cdots n_k}\}$ is a basis for the topology on N^∞, and note that it also forms an open sieve.

The following result implies that N^∞ is a "universal" Polish space in the sense that every Polish space is homeomorphic to a quotient space of N^∞.

Theorem 3.1.4. *For every Polish space P, there is a continuous open mapping of N^∞ onto P.*

Proof. Let $\{A_{n_1 \cdots n_k}\}$ be an open sieve for P having the properties of 3.1.3. We will define a function $f: N^\infty \to P$ as follows. Choose $v = (n_1, n_2, \ldots) \in N^\infty$. By the remarks following 3.1.3, we may define $f(v)$ as the unique element of $\bigcap_{k=1}^{\infty} A_{n_1 \cdots n_k}$.

Now let $\{U_{n_1 \cdots n_k}\}$ be the system of open sets in N^∞ described above. Since $\{U_{n_1 \cdots n_k}\}$ (resp. $\{A_{n_1 \cdots n_k}\}$) forms a base for the topology of N^∞ (resp. P) and since the diameter of each $A_{n_1 \cdots n_k}$ is $\leqslant 1/k$, a few moments thought shows that all the desired properties of f (i.e., continuity openness, surjectivity) will follow provided that we prove $f(U_{n_1 \cdots n_k}) = A_{n_1 \cdots n_k}$, for every $k \geqslant 1$, $n_1, \ldots, n_k \geqslant 1$. For that, let $v \in U_{n_1 \cdots n_k}$. Then v has the form $v = (n_1, \ldots, n_k, n_{k+1}, \ldots)$ and so $f(v) \in \bigcap_{j=1}^{\infty} A_{n_1 \cdots n_j} \subseteq A_{n_1 \cdots n_k}$, proving one inclusion. Conversely, if $x \in A_{n_1 \cdots n_k}$, then $A_{n_1 \cdots n_k} = \bigcup_{\ell=1}^{\infty} A_{n_1 \cdots n_k \ell}$ implies that there is an n_{k+1} so that $x \in A_{n_1 \cdots n_{k+1}}$. Repeating the argument on $A_{n_1 \cdots n_{k+1}}$ gives n_{k+2} so that $x \in A_{n_1 \cdots n_{k+2}}$, and so on inductively. Thus we can find a sequence $n_{k+1}, n_{k+2}, \ldots$ so that $x \in A_{n_1 \cdots n_j}$ for $j \geqslant k$. This implies $x \in \bigcap_{k=1}^{\infty} A_{n_1 \cdots n_k}$ (recall that the sequence $E_k = A_{n_1 \cdots n_k}$ is decreasing), and therefore $x = f(v)$ where $v = (n_1, n_2, \ldots)$. Since $v \in U_{n_1 \cdots n_k}$, we are done. □

3.2. Borel Sets and Analytic Sets

Let P be a Polish space. A *Borel set* in P is a member of the σ-field generated by the closed subsets of P. If f is a continuous map of P into another Polish space Q, then while the inverse images under f of closed sets are always closed, this is of course not so for direct images, even if f is one-to-one. Indeed $f(P)$ is in general not even a Borel set in Q. A remarkable theorem of Souslin, however, implies that a *one-to-one* continuous function does map closed sets to Borel sets (and therefore Borel sets to Borel sets). Thus it is not too surprising that this result has great significance in the theory of Borel structures. We will give a proof of Souslin's theorem in this section, and in the next we will extract numerous consequences from it.

A subset of a Polish space P is called *analytic* if it has the form $f(Q)$, where Q is a Polish space and f is a continuous map of Q into P. First, we want to show that Borel sets are analytic sets of a special kind.

Lemma. *Let $\mathscr{F}$ be a family of subsets of a set S, containing S and $\varnothing$, and which is closed under countable intersections and countable disjoint unions. Let $\mathscr{F}_0 = \{E \in \mathscr{F} : S\backslash E \in \mathscr{F}\}$. Then $\mathscr{F}_0$ is a σ-field.*

Proof. Clearly $\varnothing \in \mathscr{F}_0$, and $\mathscr{F}_0$ is closed under complementation. So $\mathscr{F}_0$ will be a σ-field provided we show that it is closed under countable intersections. But if $E_n \in \mathscr{F}_0$ then both E_n and $S\backslash E_n$ belong to $\mathscr{F}$, so that $\bigcap_n E_n \in \mathscr{F}$, and $S\backslash\bigcap_n E_n = \bigcup_n (S\backslash E_n) = \bigcup_n (S\backslash E_n \cap E_{n-1} \cap \cdots \cap E_2 \cap E_1)$ is a countable disjoint union of elements of $\mathscr{F}$. Thus $S\backslash\bigcap_n E_n \in \mathscr{F}$, and hence $\bigcap_n E_n \in \mathscr{F}_0$. □

Theorem 3.2.1. *Let P be a Polish space. Then for every Borel set $E \subseteq P$, there is a Polish space Q and a 1—1 continuous function $f: Q \to P$ such that $E = f(Q)$.*

Proof. Let $\mathscr{F}$ be the class of all subsets of P having the stated property. We will show that family of all subsets E, such that *both* E and $P\backslash E$ belong to $\mathscr{F}$, contains the Borel sets. For that, according to the preceding lemma, it suffices to show that $\mathscr{F}$ contains the open sets, the closed sets, and is closed under countable intersections and countable disjoint unions. It is clear that $\mathscr{F}$ contains the open sets and closed sets since they are themselves Polish subspaces.

Now let $E_1, E_2, \ldots$ belong to $\mathscr{F}$. We claim: $\bigcap_n E_n \in \mathscr{F}$. Choose Polish spaces P_n and continuous 1—1 functions $f_n: P_n \to P$ such that $f_n(P_n) = E_n$. Define a function $f: \prod_n P_n \to P \times P \times \cdots$ by $f(p_1, p_2, \ldots) = (f_1(p_1), f_2(p_2), \ldots)$. Then f is obviously continuous and 1—1. Since the set $\Delta = \{(x_1, x_2, \ldots) \in P \times P \times \cdots : x_1 = x_2 = x_3 = \cdots\}$ is closed, $Q = f^{-1}(\Delta)$ is a closed subspace of the domain of f, and is therefore Polish. Note that f is an injection of Q onto $\{(x, x, x, \ldots): x \in \bigcap_n E_n\}$, so that if p_1 is the projection of $P \times P \times \cdots$ onto its first coordinate space, then the composition $p_1 \circ f|_\Delta$ is the required continuous 1—1 map of Q onto $\bigcap_n E_n \in \mathscr{F}$.

Finally, if $E_n \in \mathscr{F}$ and $E_m \cap E_n = \varnothing$ for $m \neq n$, then choosing P_n and $f_n: P_n \to P$ as above, form the Polish space $Q = \sum_n P_n$ and define $f: Q \to P$ by $f(x) = f_n(x)$ if $x \in P_n$. f is continuous, it is 1—1 because the E_n are disjoint, and $f(Q) = \bigcup_n E_n$. That shows $\bigcup_n E_n \in \mathscr{F}$. □

This theorem is a slight weakening of a result of Souslin's, which asserts that every Borel set is the continuous 1—1 image of a zero-dimensional Polish space ([19], p. 447).

Corollary. *Every Borel set in a Polish space is analytic.*

The converse of this corollary is false (c.f. [19], p. 479, for an example).

The next result, due to Lusin ([19], p. 485), provides a key step in the proof of 3.2.3 below, and is of considerable interest by itself. First, some terminology. Two disjoint subsets A, B of a Polish space are said to be *separated* if there are disjoint Borel sets E, F such that $A \subseteq E$, $B \subseteq F$. If A_n, B_n are two sequences of subsets such that every A_m is separated from every B_n, then some elementary manipulations with countable intersections and unions (which we leave for the reader) show that $\bigcup_n A_n$ and $\bigcup_n B_n$ are separated.

Theorem 3.2.2 Separation theorem. *Two disjoint analytic sets in a Polish space are separated.*

Proof. Let P be a Polish space and let A and B be disjoint analytic subsets of P. By 3.1.4, every analytic set is the continuous image of N^∞, so there are continuous functions f, $g: N^\infty \to P$ such that $A = f(N^\infty)$, $B = h(N^\infty)$.

Assume now that A and B are not separated. Let $\{U_{n_1 \cdots n_k}\}$ be the basis for N^∞ described in the discussion preceding 3.1.4. Since $N^\infty = \bigcup_n U_n$ we have $A = \bigcup_n f(U_n)$ and $B = \bigcup_n g(U_n)$, so by the preceding remarks there exist integers m_1, n_1 such that $f(U_{m_1})$ and $g(U_{n_1})$ are not separated. Similarly $U_{m_1} = \bigcup_\ell U_{m_1\ell}$ and $U_{n_1} = \bigcup_\ell U_{n_1\ell}$, so there exist m_2, n_2 so that $f(U_{m_1m_2})$ and $g(U_{n_1n_2})$ are not separated. Continuing inductively, we find $m_1, m_2, \ldots$, $n_1, n_2, \ldots$ such that $f(U_{m_1 \cdots m_k})$ and $g(U_{n_1 \cdots n_k})$ are not separated for every $k \geqslant 1$. Define points $\mu, \nu \in N^\infty$ by $\mu = (m_1, m_2, \ldots)$, $\nu = (n_1, n_2, \ldots)$. Now $f(\mu) \neq g(\nu)$ because A and B are disjoint, so there exist disjoint open sets

U, V such that $f(\mu) \in U$ and $g(\nu) \in V$. Now for each k, $U_{m_1 \cdots m_k}$ is a neighborhood of μ for which diam $U_{m_1 \cdots m_k} \to 0$ as $k \to \infty$. Thus by continuity we have $f(U_{m_1 \cdots m_k}) \subseteq U$ for large enough k. Similarly $g(U_{n_1 \cdots n_k}) \subseteq V$ for large k. But this contradicts the fact that by construction $f(U_{m_1 \cdots m_k})$ is not separated from $g(U_{n_1 \cdots n_k})$. □

Note the close parallel between this argument and the customary proof of the Bolzano–Weierstrass theorem.

Corollary 1. *Let S be a subset of a Polish space P. If both S and $P \backslash S$ are analytic, then S is a Borel set.*

We can already deduce a weak form of Souslin's theorem (3.2.3 below):

Corollary 2. *Let P and Q be Polish spaces, and let f be a continuous 1–1 function of P onto Q. Then f maps closed sets to Borel sets (and therefore Borel sets to Borel sets).*

PROOF. Let $F \subseteq P$ be closed. Because f is 1—1 and onto, $f(F)$ and $f(P \backslash F)$ are complementary sets in Q. Since F and $P \backslash F$ are Polish subspaces (cf. 3.1.1) and since f is continuous, $f(F)$ and $f(P \backslash F)$ are both analytic sets. An application of Corollary 1 finishes the proof. □

The next corollary follows from 3.2.2 by routine manipulations with countable unions and intersections, which we omit.

Corollary 3. *Let A_n be a disjoint sequence of analytic subsets of a Polish space. Then there are disjoint Borel sets E_n such that $A_n \subseteq E_n$, for every n.*

Before proceeding with the main theorem, we need some preliminaries. A *partition* of a set S is a family $\mathscr{P}$ of disjoint subsets of S for which $\bigcup \mathscr{P} = S$. A partition $\mathscr{P}_2$ is said to be a *refinement* of a partition $\mathscr{P}_1$ (this is written $\mathscr{P}_2 \leqslant \mathscr{P}_1$) if each set in $\mathscr{P}_2$ is contained in some set in $\mathscr{P}_1$. Note that this is equivalent to requiring that every set in $\mathscr{P}_1$ be a union of sets in $\mathscr{P}_2$. We want to point out now that every Polish space P contains a decreasing sequence $\mathscr{P}_1 \geqslant \mathscr{P}_2 \geqslant \cdots$ of *countable* partitions such that each element of $\mathscr{P}_n$ is a G_δ whose diameter (relative to a fixed metric on P) is at most $1/n$. Indeed, as in the proof of 3.1.3, we can cover P with a sequence $B_1, B_2, \ldots$ of open balls having diameter $\leqslant 1$. Let $\mathscr{P}_1 = \{A_1, A_2, \ldots\}$, where $A_1 = B_1$, $A_n = B_n \backslash (B_1 \cup \cdots \cup B_{n-1})$, $n > 1$. Each A_n is a G_δ, being an intersection of an open set with a closed set, and thus we have $\mathscr{P}_1$. Inductively, given $\mathscr{P}_1, \ldots, \mathscr{P}_n$ as above, choose a sequence $C_1, C_2, \ldots$ of open sets having diameter $\leqslant 1/n + 1$ such that $\bigcup_n C_n = P$, and put $D_1 = C_1$, $D_n = C_n \backslash (C_1 \cup \cdots \cup C_{n-1})$, $n \geqslant 1$. Then $\mathscr{P}_{n+1} = \{A \cap D_k : A \in \mathscr{P}_n,\ k = 1, 2, \ldots\}$ has all the stated properties (note for example that $A \cap D$ is a G_δ when both A and D are).

Theorem 3.2.3. *Let P, Q be Polish spaces and let $f: P \to Q$ be a 1—1 continuous function. Then $f(P)$ is a Borel set in Q.*

PROOF. Choose a sequence $\mathscr{P}_1 \geqslant \mathscr{P}_2 \geqslant \cdots$ of countable partitions of P as in the preceding remark, say $\mathscr{P}_n = \{A_{n1}, A_{n2}, \ldots\}$, $n \geqslant 1$. Now since each set A_{nk} is a G_δ it is a Polish subspace of P. Thus for each n, $\{f(A_{n1}), f(A_{n2}), \ldots\}$ forms a disjoint sequence of analytic sets in Q, and by Corollary 3 above there are disjoint Borel sets B_{n1}, $B_{n2}, \ldots$ such that $f(A_{nk}) \subseteq B_{nk}$, $k = 1, 2, \ldots$. By replacing B_{nk} with $B_{nk} \cap f(A_{nk})^-$ we can assume $f(A_{nk}) \subseteq B_{nk} \subseteq f(A_{nk})^-$. Moreover, by intersecting each element of $\{B_{n+1\,1}, B_{n+1\,2}, \ldots\}$ with an appropriate element of $B_{n1}, B_{n2}, \ldots\}$ (utilizing finite induction on n) we can assume that each nonvoid $B_{n+1\,k}$ is contained in a unique member of $\{B_{n1}, B_{n2}, \ldots\}$, $n \geqslant 1$. Note that this implies $A_{n+1\,k} \subseteq A_{nj}$ if, and only if, $B_{n+1\,k} \subseteq B_{nj}$, for all $n, j, k \geqslant 1$.

We will show that $f(P) = \bigcap_{n=1}^{\infty} \bigcup_{k=1}^{\infty} B_{nk}$. This will complete the proof, for the right side is clearly a Borel set. The inclusion $\subseteq$ is obvious, so choose a point q in the right hand member. Then for each n there is a k_n so that $q \in B_{nk_n}$. Thus $B_{nk_n} \cap B_{n+1\,k_{n+1}} \neq \varnothing$ and so $B_{n+1\,k_{n+1}} \subseteq B_{nk_n}$. By the preceding paragraph we have $A_{n+1\,k_{n+1}} \subseteq A_{nk_n}$. Also $q \in B_{nk_n} \subseteq f(A_{nk_n})^-$. Thus $\bar{A}_{1k_1}, \bar{A}_{2k_2}, \ldots$ forms a decreasing sequence of nonvoid closed sets in P whose diameters shrink to 0. By the Cantor nested sets property they intersect in a unique point p. To see that $q = f(p)$, let V be any closed neighborhood of $f(p)$. By continuity at p and the fact that diam $\bar{A}_{nk_n} \to 0$ we will have $f(A_{nk_n}) \subseteq V$ for large enough n, and therefore $q \in B_{nk_n} \subseteq f(A_{nk_n})^- \subseteq V$ for large n. Since V was arbitrary this means $q = f(p)$. □

Corollary. *Let $f: P \to Q$ satisfy the hypotheses of the preceding theorem. Then f maps Borel sets to Borel sets.*

PROOF. 3.2.3 implies that f maps closed sets to Borel sets; since f is 1—1 and the closed sets in P generate its Borel structure, the conclusion is immediate. □

The final result in this section is a measurability theorem for analytic sets, adapted from ([19], p. 482). A *finite Borel measure* on a Polish space P is a mapping μ from the Borel sets of P into $[0, \infty)$ such that $\mu(\varnothing) = 0$ and $\mu(\bigcup_n E_n) = \sum_n \mu(E_n)$ for every sequence $E_1, E_2, \ldots$ of disjoint Borel sets. A subset $S \subseteq P$ is *μ-measurable* if there exist Borel sets $B_1 \subseteq S \subseteq B_2$ for which $\mu(B_2 \backslash B_1) = 0$. This is equivalent to the existence of Borel sets B, N such that $S \subseteq B$ and $B \backslash S \subseteq N$, where $\mu(N) = 0$.

Theorem 3.2.4. *Let A be an analytic set in a Polish space P. Then A is μ-measurable for every finite Borel measure μ on P.*

PROOF. Fix μ, a finite Borel measure, and let S be an arbitrary subset of P. A Borel set E containing S will be called minimal if for every other Borel set $F \supseteq S$ one has $\mu(E\backslash F) = 0$. Note that every subset S has a minimal Borel cover; for if we choose Borel sets $E_n \supseteq S$ so that $\mu(E_n)$ tends to $\alpha = \inf\{\mu(F)$: F Borel, $F \supseteq S\}$, then $E = \bigcap_n E_n$ is a Borel cover of S for which $\mu(E) = \alpha$, and thus it is minimal because for every Borel set $F \supseteq S$, $E \cap F \supseteq S$ implies $\mu(E \cap F) \geqslant \alpha$, hence $\mu(E\backslash F) = \mu(E) - \mu(E \cap F) \leqslant \mu(E) - \alpha = 0$.

Now let A be an analytic set in P. We will produce Borel sets B, N such that $A \subseteq B$, $B\backslash A \subseteq N$, and $\mu(N) = 0$. As in the proof of 3.2.2 there is a continuous function $f: N^\infty \to P$ such that $A = f(N^\infty)$. Let $\{U_{n_1 \cdots n_k}\}$ be the basis for N^∞ described in the discussion preceding 3.1.4. For each $k \geqslant 1$ and $n_1, \ldots, n_k \geqslant 1$, choose a minimal Borel cover $E_{n_1 \cdots n_k}$ for $f(U_{n_1 \cdots n_k})$. By intersecting with $f(U_{n_1 \cdots n_k})^-$ if necessary, we can assume

$$f(U_{n_1 \cdots n_k}) \subseteq E_{n_1 \cdots n_k} \subseteq f(U_{n_1 \cdots n_k})^-.$$

Define $B = \bigcup_{n_1=1}^\infty E_{n_1}$, and define

$$N = \bigcup_{k=1}^\infty \bigcup_{n_1, \ldots, n_k=1}^\infty \left(E_{n_1 \cdots n_k} \middle\backslash \bigcup_{\ell=1}^\infty E_{n_1 \cdots n_k \ell} \right).$$

It is clear that both B and N are Borel sets, and $A \subseteq B$ is immediate from $A = f(N^\infty) = \bigcup_{n_1} f(U_{n_1}) \subseteq B$. Thus we must prove $\mu(N) = 0$ and $B\backslash A \subseteq N$.

Now since $U_{n_1 \cdots n_k} = \bigcup_{\ell=1}^\infty U_{n_1 \cdots n_k \ell}$ we have

$$f(U_{n_1 \cdots n_k}) = \bigcup_{\ell=1}^\infty f(U_{n_1 \cdots n_k \ell}) \subseteq \bigcup_{\ell=1}^\infty E_{n_1 \cdots n_k \ell}.$$

By minimality we conclude that $\mu(E_{n_1 \cdots n_k} \backslash \bigcup_{\ell=1}^\infty E_{n_1 \cdots n_k \ell}) = 0$ for every $k \geqslant 1$, $n_1, \ldots, n_k \geqslant 1$, and thus N, being the countable union of sets of measure zero, has measure zero. To prove $B\backslash A \subseteq N$, choose $p \in B\backslash A$. Then there exists $n_1 \geqslant 1$ such that $p \in E_{n_1}\backslash A$. Assume, contrapositively, that $p \notin N$. Then in particular $p \notin E_{n_1}\backslash \bigcup_{\ell=1}^\infty E_{n_1 \ell}$, and since $p \in E_{n_1}$ we can only have $p \in \bigcup_{\ell=1}^\infty E_{n_1 \ell}$. Thus there is an n_2 such that $p \in E_{n_1 n_2}$. Similarly $p \notin E_{n_1 n_2}\backslash \bigcup_{\ell=1}^\infty E_{n_1 n_2 \ell}$ (because $p \notin N$) and since $p \in E_{n_1 n_2}$ there must be an n_3 so that $p \in E_{n_1 n_2 n_3}$. Continuing in this way we obtain a sequence $n_1, n_2, \ldots$ of positive integers such that $p \in E_{n_1 \cdots n_k}$ for every $k \geqslant 1$. Define $v \in N^\infty$ by $v = (n_1, n_2, \ldots)$. Now for each k $U_{n_1 \cdots n_k}$ is a neighborhood of v for which diam $U_{n_1 \cdots n_k} \to 0$ as $k \to \infty$. Thus by continuity at v we have $\bigcap_{k=1}^\infty f(U_{n_1 \cdots n_k})^- = \{f(v)\}$. Since $p \in E_{n_1 \cdots n_k} \subseteq f(U_{n_1 \cdots n_k})^-$ for every k, it follows that $p = f(v) \in A$, and this contradiction completes the proof. □

Thus while analytic sets are not necessarily Borelian, at worst they differ from Borel sets in a metrically trivial way. Note also that the preceding theorem implies that a good many nonanalytic sets are measurable for every finite measure μ. Indeed since the μ-measurable subsets of P form a σ-field, it follows that every set in the σ-field generated by the analytic sets is measurable. In particular the complement of an analytic set A is measurable, while by Corollary 1 of 3.2.2 $P\backslash A$ will *not* be analytic if A is not a Borel set.

Finally, observe that this theorem is also true for σ-finite measures, since every σ-finite measure is mutually absolutely continuous with a finite measure.

3.3. Borel Spaces

In this section we deduce some rather striking consequences of the preceding material. Most of these results are due to Mackey [20]. A *Borel space* is a pair $(X, \mathscr{B})$ where X is a set and $\mathscr{B}$ is a σ-field of subsets of X, whose elements we will call Borel sets. Thus every topological space X gives rise to a natural Borel structure $\mathscr{B}$, the σ-field generated by the closed subsets of X. Unless there is cause for confusion, we will suppress the letter $\mathscr{B}$ in the notation for Borel spaces. A *subspace* of a Borel space X is a subset $S \subseteq X$ endowed with the relative Borel structure, namely the σ-field of all subsets of S of the form $S \cap E$ where E is a Borel subset of X. A mapping $f: X \to Y$ between Borel spaces X and Y is a *Borel function* if $f^{-1}(E)$ is a Borel set in X for every Borel set $E \subseteq Y$. f is a *Borel isomorphism* if it is 1—1, onto, and both f and f^{-1} are Borel functions (we have resisted an impulse to use the term *Borelomorphism*). There are natural definitions for the Cartesian product and direct sum of families of Borel spaces; for example if $X_1, X_2, \ldots$ is a sequence of Borel spaces then the Borel structure in $\prod_n X_n$ is the σ-field generated by all finite "rectangles" of the form $\{(x_1, x_2, \ldots): x_1 \in E_1, \ldots, x_n \in E_n\}$, where E_i is an arbitrary Borel subset of X_i, $1 \leqslant i \leqslant n$, $n = 1, 2, \ldots$. Finally, a Borel space X is *countably separated* if there is a sequence $E_1, E_2, \ldots$ of Borel subsets of X which separates points of X (i.e., if $x \neq y$ then there is an n such that $\chi_{E_n}(x) \neq \chi_{E_n}(y)$, χ_E denoting the characteristic function of E).

We begin with a simple and useful lemma.

Proposition 3.3.1. *Let X, Y be Borel spaces, with Y countably separated, and let $f: X \to Y$ be a Borel function. Then the graph of f is a Borel subset of $X \times Y$.*

PROOF. Let $E_1, E_2, \ldots$ be a separating family of Borel sets in Y. Let $\Delta = \{(y, y): y \in Y\}$ be the diagonal in $Y \times Y$. Then $(Y \times Y)\backslash\Delta$ is the union of the two sets $\bigcup_n E_n \times (Y\backslash E_n)$ and $\bigcup_n (Y\backslash E_n) \times E_n$, so that Δ is a Borel subset of $Y \times Y$. Now the function $\varphi: X \times Y \to Y \times Y$ defined by $\varphi(x, y) = (f(x), y)$ is Borel (since each coordinate is a Borel function), and since the graph of f is simply $\varphi^{-1}(\Delta)$, the proof is complete. □

A Borel space X is called *standard* if X is isomorphic to a Borel subset of a Polish space. A noteworthy fact, which we do not need, is that every uncountable standard Borel space is isomorphic to the unit interval $[0, 1]$ with its usual Borel structure ([19], remark (i), p. 451). Thus a great variety of mathematical constructions lead to identical Borel spaces. Nevertheless, as we will see presently, standard Borel spaces do not quite suffice for dealing with problems in analysis involving equivalence relations.

The following generalizes Souslin's theorem (Theorem 3.2.3).

Theorem 3.3.2. *Let X be a standard Borel space, let Q be a Polish space, and let f be a 1—1 Borel map of X into Q. Then $f(X)$ is a Borel set in Q and f is an isomorphism of X onto $f(X)$.*

PROOF. We will show that $f(X)$ is a Borel set; the rest follows from this and the obvious fact that a Borel subset of the standard Borel space X defines a standard subspace.

By definition, we may assume that X is a Borel subset of a Polish space P_0; and by 3.2.1. there is a 1—1 continuous map g of a Polish space P onto X. Thus $f(X) = f \circ g(P)$ and $f \circ g$ is 1—1 Borel function from P into Q. Hence we may even assume that $X = P$ is a Polish space.

Now Q is countably separated (take $E_1, E_2, \ldots$ to be a countable base for the topology of Q) so that by 3.3.1. the graph of f is a Borel set in the Polish space $P \times Q$. Utilizing 3.2.1. again, we can find a Polish space R and a 1—1 continuous function $h: R \to P \times Q$ such that $h(R) = \text{graph } f$. The projection p_2 of $P \times Q$ onto Q is continuous, its restriction to graph f is 1—1 (since f is 1—1), and thus $p_2 \circ h$ is a continuous 1—1 function from R into Q whose range is $f(P)$. By Souslin's theorem 3.2.3 $f(P)$ is a Borel set. □

Let X be a Borel space and let $\sim$ be an equivalence relation in X. Let $X/\sim$ be the set of all equivalence classes. Then the function $q: X \to X/\sim$ which assigns to each point of X the equivalence class containing it maps onto $X/\sim$, and thus determines a Borel structure on $X/\sim$: by definition a subset E of $X/\sim$ is a Borel set if $q^{-1}(E)$ is a Borel set in X. As an alternate description, let us say a Borel subset $B \subseteq X$ is *saturated* if B is a union of equivalence classes, i.e., $x \in B$ and $y \sim x$ implies $y \in B$. Then $E \leftrightarrow q^{-1}(E)$ is a bijective correspondence between the Borel sets in $X/\sim$ and the saturated Borel sets in X. Note that q is a Borel map of X onto $X/\sim$. Here is a simple example illustrating the kind of equivalence relation one encounters.

Example 3.3.3. Consider the classical problem of classifying complex irreducible $n \times n$ matrices with respect to unitary equivalence. Fix n, a positive integer, let M_n denote the set of all complex $n \times n$ matrices, and let $\mathscr{I}_n$ be the set of irreducible members of M_n (a matrix is irreducible if it commutes with no self-adjoint projections other than 0 and the identity). M_n is a Polish space in its natural topology (for example, the distance between two matrices (a_{ij}) and (b_{ij}) can be taken as $\sum_{i,j} |a_{ij} - b_{ij}|$ and it is not very hard to see that $\mathscr{I}_n$ is a G_δ in M_n. Indeed, $x \in M_n$ is irreducible if the C^*-algebra generated by x is all of M_n and, choosing x_0 to be a fixed irreducible element of M_n, this is true iff the C^*-algebra generated by x contains x_0.

If we let $P_1(\xi, \eta)$, $P_2(\xi, \eta)$, ... be the countable set of all noncommutative polynomials in the two free variables ξ, η having rational coefficients, then for each $i \geq 1$ the function $x \mapsto p_i(x, x^*)$ is continuous, hence each set $G_j = \bigcup_{i=1}^{\infty} \{x \in M_n : \text{dist}(x_0, p_i(x, x^*)) < 1/j\}$ is open, and so $\mathscr{I}_n = \bigcap_{j=1}^{\infty} G_j$ is a G_δ. By 3.1.2, $\mathscr{I}_n$ is a Polish space. Define $\sim$ in $\mathscr{I}_n$ by $x \sim y$ iff x and y are unitarily equivalent. Thus the quotient Borel space $\mathscr{I}_n/\sim$ can be regarded as a *classification space* for irreducible $n \times n$ matrices. Now the classification problem, that is, the problem of finding a complete set of unitary invariants for $\mathscr{I}_n$, amounts to finding a convenient parametrization of $\mathscr{I}_n/\sim$ in terms of the familiar spaces of analysis. More precisely, one wants a countable set of "well-defined" (i.e., Borel) functions from $\mathscr{I}_n/\sim$ into, say, the real or complex numbers (or in general some Polish space), which separates points in $\mathscr{I}_n/\sim$. The existence of such a family of functions is easily seen to be equivalent to saying $\mathscr{I}_n/\sim$ is countably separated, and thus a knowledge of the Borel structure of $\mathscr{I}_n/\sim$ is of key importance. We shall return to this example after Corollary 2 of 3.3.5 below.

It often happens that the quotient of a standard Borel space by a very regular equivalence relation fails to be standard. The relevant definition is the following. A Borel space X is called *analytic* if there is an analytic set A in a Polish space such that X is isomorphic to A with its relative Borel structure. Note that by the corollary of 3.2.1 each standard Borel space is analytic. The converse is, however, false; see ([19], p. 479) for some interesting examples. Thus while there is only one uncountable standard Borel space modulo isomorphism, this is not true of uncountable analytic Borel spaces. So far as we know, the latter have not been classified. We now prove the key invariance property of analytic Borel spaces.

Theorem 3.3.4. *Let X be an analytic Borel space, let Q be a Polish space, and let f be a Borel map of X into Q. Then $f(X)$ is an analytic set in Q.*

PROOF. The argument is similar to that of 3.3.2. We can assume X is an analytic subspace of a Polish space P_0. Then by definition there is a continuous map of a Polish space P into P_0 having range X. Thus we may assume (by considering the composition of the two functions) that $X = P$ is itself Polish.

By 3.3.1 the graph of f is a Borel set in $P \times Q$, and so there is a continuous (in fact, 1—1) function g from a Polish space R into $P \times Q$ having range graph f. Letting p_2 be the projection of $P \times Q$ onto Q, we see that $p_2 \circ g$: $R \to Q$ is continuous and has range $f(P)$. Thus $f(P)$ is analytic. □

The next result shows that the analytic subspaces of Polish spaces are what you think they are.

Corollary 1. *Let X be a subspace of a Polish space P. Then X is an analytic Borel space in its relative structure if, and only if, X is an analytic set in P.*

PROOF. The if part is obvious. Conversely, suppose there exists an analytic set Y in a Polish space Q and a Borel isomorphism f of X on Y. Now apply 3.3.4 to f^{-1} to conclude that X is an analytic set. □

We remark that Theorem 3.3.2 has a corollary analogous to Corollary 1; the proof is the same.

Corollary 2. *Let X and Y be analytic Borel spaces and let f be a 1—1 Borel map of X onto Y. Then f is a Borel isomorphism.*

PROOF. Again we may assume X and Y are analytic subspaces of Polish spaces P and Q. Let E be a Borel set in X; we have to show that $f(E)$ is a Borel set in Y. E has the form $X \cap B$ for some Borel set $B \subseteq P$. Thus E is analytic (here we use the fact that a countable intersection of analytic sets is analytic, the proof of which is a trivial adaptation of an argument in the proof of 3.2.1), so by 3.3.4 $f(E)$ is an analytic set in Q. Similarly $f(X\backslash E)$ is analytic, and $f(E) \cap f(X\backslash E) = \emptyset$ because f is 1—1. By the separation theorem (Theorem 3.2.2) there is a Borel set $C \subseteq Q$ such that $f(E) \subseteq C$ and $f(X\backslash E) \subseteq Q\backslash C$, which shows that $f(E) = f(X) \cap C = Y \cap C$ is a Borel set in Y. □

We remark that Corollary 2 remains valid if we replace the hypothesis that Y be analytic with the weaker requirement that Y be countably separated. See Corollary 1 of 3.3.5. The next result is an elegant Borel space analogue of the Stone–Weierstrass theorem. First, we need a lemma. A Borel space $(X, \mathscr{B})$ is said to be *countably generated* if there is a countable subset of $\mathscr{B}$ which separates points and generates $\mathscr{B}$ as a σ-field. It is easy to check that *every* generating subfamily of $\mathscr{B}$ must separate points of X.

Lemma. *A Borel space is isomorphic to a subspace of a Polish space if, and only if, it is countably generated.*

PROOF. The only if part is trivial, since every Polish space is countably generated (consider a countable base for the topology) and hence so is every subspace of it. Conversely suppose $E_1, E_2, \ldots$ are a generating family for the Borel structure on X. Define P to be the cartesian product $\prod_n P_n$, where P_n is the discrete two-element set $\{0, 1\}$, $n = 1, 2, \ldots$. Then P is a Polish space and we have a map $\varphi: X \to P$ defined by $\varphi(x) = (\chi_{E_1}(x), \chi_{E_2}(x), \ldots)$, χ_E denoting the characteristic function of E. φ is 1—1 because $\{E_n\}$ separates points. Moreover if we define $F_n \subseteq P$ by $F_n = \{(x_1, x_2, \ldots) \in P: x_n = 1\}$, then $\{F_1, F_2, \ldots\}$ generates the Borel structure in P, so that $\{\varphi(X) \cap F_n\}$ generates the Borel structure in $\varphi(X)$. Finally, since $\varphi(E_j) = \varphi(X) \cap F_j$, it follows that φ is a Borel isomorphism of X on $\varphi(X)$. □

Theorem 3.3.5. Unique structure theorem. *Let $(X, \mathscr{B})$ be an analytic Borel space and let $\mathscr{B}_0$ be a countably generated sub-σ-field of $\mathscr{B}$ which separates points in X. Then $\mathscr{B}_0 = \mathscr{B}$.*

Proof. By the lemma, there is a Borel isomorphism f of $(X, \mathscr{B}_0)$ onto a subspace Y of a Polish space P. Note that the Borel structure on Y is $f(\mathscr{B}_0)$. Now f is also a 1—1 Borel map of $(X, \mathscr{B})$ into $(Y, f(\mathscr{B}_0))$ (because $\mathscr{B}$ contains $\mathscr{B}_0$), so by Theorem 3.3.4 Y is an analytic set in P, and by Corollary 2 f is an isomorphism of $(X, \mathscr{B})$ on $(Y, f(\mathscr{B}_0))$. In particular, $\mathscr{B} = f^{-1}(f(\mathscr{B}_0)) = \mathscr{B}_0$, completing the proof. □

We remark that the preceding result is false without the hypothesis that $\mathscr{B}_0$ be countably generated. Here is a simple example. Let $(X, \mathscr{B})$ be the standard Borel space consisting of the unit interval $[0, 1]$ with its usual Borel structure and let $\mathscr{B}_0$ be the sub σ-field of all countable subsets of $[0, 1]$ together with their complements. Clearly $\mathscr{B}_0$ separates points, but of course $\mathscr{B}_0 \neq \mathscr{B}$. Evidently, $\mathscr{B}_0$ is not countably generated and in fact no countable subfamily of $\mathscr{B}_0$ can separate points.

The next result implies that while the range of a standard Borel space under a Borel map may fail to be standard, it is often no worse than analytic.

Corollary 1. *Let X be an analytic Borel space, let Y be a countably separated Borel space, and let f be a Borel map of X onto Y. Then Y is an analytic Borel space.*

Proof. Let $\{E_1, E_2, \ldots\}$ be a separating family of Borel sets in Y. We claim: $\{E_1, E_2, \ldots\}$ generates the Borel structure in Y. For this choose an arbitrary Borel set F in Y, and let $\mathscr{B}_0$ be the σ-field generated by $\{F, E_1, E_2, \ldots\}$. Since $(Y, \mathscr{B}_0)$ is countably generated we may by the preceding lemma regard $(Y, \mathscr{B}_0)$ as a subspace of a Polish space P. By 3.3.4 Y is an analytic set in P, thus $(Y, \mathscr{B}_0)$ is an analytic Borel space, thus by the unique structure theorem $\mathscr{B}_0$ is generated by $\{E_1, E_2, \ldots\}$. Since F was arbitrary, this proves the claim.

Thus Y is countably generated, and so the above argument may be repeated to show that Y is analytic. □

We may apply this to equivalence relations as follows.

Corollary 2. *Let X be an analytic Borel space and let $\sim$ be an equivalence relation in X. Assume there is a sequence $f_1, f_2, \ldots$ of real valued Borel functions on X such that for any pair of points x, y in X one has $x \sim y$ iff $f_n(x) = f_n(y)$ for all n. Then $X/\sim$ is an analytic Borel space.*

Proof. Let $q: X \to X/\sim$ be the natural projection. Define $g_n: X/\sim \to \mathbb{R}$ by $g_n(q(x)) = f_n(x)$. Then each g_n is a Borel function and the g's separate points in $X/\sim$. So if we let $r_1, r_2, \ldots$ be the rational numbers in $\mathbb{R}$, then the family of sets $E_{mn} = \{z \in X/\sim : g_n(z) < r_m\}$ separate, and the conclusion follows from Corollary 1. □

Note, finally, that the converse of Corollary 2 is valid, since every analytic Borel space is countably separated.

Let us return to Example 3.3.3 and examine the Borel structure of the classification space $\mathscr{I}_n/\sim$. As an application of the preceding corollary we will show that $\mathscr{I}_n/\sim$ is analytic (in fact, it is standard, as 3.4.1 and 3.4.2 in the next section will show). Let $\mathscr{W}$ be the set of all finite words $w(\xi, \eta)$ in the two free variables ξ, η; thus $w(\xi, \eta) = \xi^2\eta^3\xi\eta$ is a typical element of $\mathscr{W}$. Let tr denote the canonical trace on M_n. For each $w(\xi, \eta) \in \mathscr{W}$ define a complex valued continuous function $f_w : \mathscr{I}_n \to \mathbb{C}$ by $f_w(x) = \operatorname{tr}(w(x, x^*))$. Clearly $\{f_w : w \in \mathscr{W}\}$ is countable, and each f_w is constant on equivalence classes because the trace is unitarily invariant. So the set of all real and imaginary parts of the functions f_w will satisfy the requirements of the corollary provided we show that $f_w(x) = f_w(y)$ for all $w \in \mathscr{W}$ implies $x \sim y$, for all $x, y \in \mathscr{I}_n$. Fix x, y and assume the condition is satisfied. Then by taking linear combinations we have $\operatorname{tr}(p(x, x^*)) = \operatorname{tr}(p(y, y^*))$ for every noncommutative polynomial $p(\xi, \eta)$ in the two free variables ξ and η. We claim that the map $\pi : p(x, x^*) \mapsto p(y, y^*)$ is well-defined. For if p is a polynomial for which $p(x, x^*) = 0$, then $p(x, x^*)^*p(x, x^*) = 0$, and since $p(x, x^*)^*p(x, x^*)$ is another polynomial in x, x^* we have $\operatorname{tr}(p(y, y^*)^*p(y, y^*)) = \operatorname{tr}(p(x, x^*)^*p(x, x^*)) = \operatorname{tr}(0) = 0$, and hence $p(y, y^*) = 0$ (because $\operatorname{tr}(z^*z) = 0$ implies $z = 0$). By symmetry π is also 1—1. Note also that π preserves the algebraic operations and the $*$-operation. Thus π is a $*$-isomorphism between the C^*-algebras generated by x and by y. Since both x and y generate M_n as a C^*-algebra (by irreducibility) π is in fact a $*$-automorphism of M_n. Therefore (cf. Corollary 3 of 1.4.4) there is a unitary operator $u \in M_n$ such that $\pi(z) = uzu^*$ for all z, and in particular $y = \pi(x) = uxu^*$, proving $x \sim y$. By Corollary 2 above, $\mathscr{I}_n/\sim$ is an analytic Borel space.

Note in particular that the functions $f_w(x) = \operatorname{tr}(w(x, x^*))$ form a complete set of unitary invariants for $\mathscr{I}_n$, a fact first proved by Specht [26]. C. Pearcy has subsequently shown that for each n one can find a finite subset of $\{f_w : w \in \mathscr{H}\}$ which separates $\mathscr{I}_n/\sim$ [22]. There is of course nothing that distinguishes this particular set of invariants; for example if we let $p_1(\xi, \eta)$, $p_2(\xi, \eta), \ldots$ be the countable set of all noncommutative polynomials in the two free variables ξ, η having rational coefficients, then an argument similar to the above shows that the sequence of continuous functions $g_n(x) = \|p_n(x, x^*)\|$ ($\|z\|$ denoting the operator norm of z, regarding z as an operator on the Hilbert space $\mathbb{C}^n$) provides another countable separating family for $\mathscr{I}_n/\sim$. Debates over the relative merits of different sets of invariants in the general case have little substance; in practice one chooses the most convenient invariant parameters one can find in context.

3.4. Cross Sections

Let X be a set and let f be a function from X onto another set Y. A *cross section* for f is a function $g : Y \to X$ such that $f \circ g$ is the identity map id_Y. Note that a cross section assigns to each point $y \in Y$ a unique element $g(y)$ in the set $f^{-1}(\{y\})$; thus the existence of cross sections in general follows from (and in fact is equivalent to) the axiom of choice. For most purposes, however, this is not good enough for dealing with problems in analysis. For example, if X and Y are topological spaces and f is continuous, one might want to find a continuous cross section. Unfortunately, this is usually impossible (consider the map $f(x) = e^{ix}$ of the real line onto the unit circle), and the best one can hope for is a Borel cross section. While even that is not always possible, there is an effective theorem, due to Dixmier, which provides the Borel cross sections required for the analysis of representations of GCR C^*-algebras. After giving a proof of Dixmier's theorem, we conclude this chapter with a discussion of the general problem. It goes without saying that none of these results depend on the axiom of choice.

We begin by taking another look at the space N^∞ (see the discussion preceding 3.1.4). Define an order relation $<$ in N^∞ by $(m_1, m_2, \ldots) < (n_1, n_2, \ldots)$ iff the first k for which $m_k \neq n_k$ satifies $m_k < n_k$ (note that this is the "dictionary" order). If we define $\mu \leqslant \nu$ to mean $\mu < \nu$ or $\mu = \nu$, then $\leqslant$ is clearly a linear ordering of N^∞. We claim that in fact this is a *well-ordering* in the sense that every nonvoid closed set in N^∞ has a smallest element. Indeed if $F \neq \emptyset$ is closed, define n_1 to be the smallest first coordinate of all points in F, define n_2 to be the smallest second coordinate of all points in F having the form $(n_1, *, *, \ldots)$ (note that this set is nonvoid), n_3 is the smallest third coordinate of all points in F of the form $(n_1, n_2, *, *, \ldots)$, and so on indefinitely. Clearly $\nu = (n_1, n_2, \ldots)$ is smaller than or equals any point in F, and ν is clearly a limit point of F; i.e., $\nu \in F$. Thus N^∞ is well-ordered by $\leqslant$.

For $\nu \in N^\infty$, let us write $(-\infty, \nu)$ for the interval $\{\mu \in N^\infty : \mu < \nu\}$ and $[\nu, +\infty)$ for its complement $\{\mu \in N^\infty : \mu \geqslant \nu\}$. We claim first that each interval $(-\infty, \nu)$ is open. Indeed if $\mu = (m_1, m_2, \ldots)$ and $\nu = (n_1, n_2, \ldots)$ satisfy $\mu < \nu$, then there is a $k \geqslant 1$ for which $m_1 = n_1, \ldots, m_{k-1} = n_{k-1}$, $m_k < n_k$. Thus all points in the open set $U_{m_1 \cdots m_k} = \{(i_1, i_2, \ldots) : i_1 = m_1, \ldots, i_k = m_k\}$ are smaller than ν, proving $(-\infty, \nu)$ is open. Second we claim that the intervals $(-\infty, \nu)$ generate the Borel structure. Since the sets of the form $U_{n_1 \cdots n_k}$ (as above) form a countable base for the topology, it suffices to show that each $U_{n_1 \cdots n_k}$ has the form $(-\infty, \mu) \cap (N^\infty \backslash (-\infty, \nu)) = (-\infty, \mu) \cap [\nu, +\infty)$. But for any integers $n_1, \ldots, n_k$, it is clear that $\mu = (n_1, \ldots, n_{k-1}, n_k + 1, 1, 1, \ldots)$ and $\nu = (n_1, \ldots, n_{k-1}, n_k, 1, 1, \ldots)$ have the required property, proving the claim.

Following is a variation of a theorem of Dixmier [9]. Its proof, however, is more related to a selection theorem of von Neumann ([21], Section 15, lemma 5).

Theorem 3.4.1. *Let P be a Polish space, let Y be a Borel space, and let f be a function from P onto Y satisfying*

(i) *f maps open sets to Borel sets;*
(ii) *the inverse image of each point of Y is a closed subset of P.*

Then f has a Borel cross section.

PROOF. Assume first that the theorem has been proved for the special case $P = N^\infty$. Then for a general P we can find a continuous open map g of N^∞ onto P (3.1.4). Now the function $f \circ g : N^\infty \to Y$ satisfies (i) because g is an open map, and it satisfies (ii) because g is continuous. Thus by assumption there is a Borel cross section h for $f \circ g$, and so $g \circ h$ provides the required Borel cross section for f.

Let us now prove the theorem for $P = N^\infty$. For each $y \in Y$, $f^{-1}(\{y\})$ is closed in N^∞, thus by the well-ordering property $f^{-1}(\{y\})$ has a smallest element denoted $h(y)$. Clearly $h: Y \to N^\infty$ is a cross section for f, and we must prove that h is Borelian. Since the open intervals $(-\infty, v)$ generate the Borel structure it suffices to prove that $h^{-1}(-\infty, v)$ is a Borel set in Y, for each $v \in N^\infty$, and by property (i) this will be true if we show $h^{-1}(-\infty, v) = f(-\infty, v)$. For $\subseteq$, note that $h^{-1}(E) \subseteq f(E)$ for every set $E \subseteq N^\infty$ (Indeed $z \in h^{-1}(E)$ implies $h(z) \in E$ implies $z = f \circ h(z) \in f(E)$, because h is a cross section). For $\supseteq$, $h \circ f(\mu)$ is the smallest element in $f^{-1}(\{f(\mu)\})$, and in particular $h \circ f(\mu) \leqslant \mu$, $\mu \in N^\infty$. Thus $h \circ f(-\infty, v) \subseteq (-\infty, v)$ and hence $f(-\infty, v) \subseteq h^{-1}(-\infty, v)$. □

Let S be a set and let G be a group of transformations of S onto itself (thus we assume G contains the identity map, that each $\gamma \in G$ is invertible, and that G is a group under composition). Then we may define an equivalence relation in S by $p \sim q$ iff there exists $\gamma \in G$ such that $\gamma(p) = q$, and the quotient $S/\sim$ is called the *orbit space* of S (mod G). The following cross section theorem is the one we need for the discussion of C^*-algebras.

Corollary. *Let P be a Polish space, let G be a group of homeomorphisms of P, and let $\sim$ be the equivalence relation determined by G. Assume that each orbit $G(p) = \{\gamma(p) : \gamma \in G\}$ is closed, $p \in P$. If $P/\sim$ has the quotient Borel structure, then the canonical projection $q: P \to P/\sim$ has a Borel cross section.*

PROOF. Taking $Y = P/\sim$ in 3.4.1, we see that $q: P \to P/\sim$ satisfies condition (ii), by hypothesis. For (i), let U be open. To see that $q(U)$ is a Borel set, we must show that $q^{-1}(q(U))$ is a Borel set in P. But the latter is simply $\bigcup_{\gamma \in G} \gamma(U)$, an open set. The conclusion is now immediate. □

The next result implies that *if* a Borel map $f: X \to Y$ has a Borel cross section, then Y can be no more pathological than X.

Proposition 3.4.2. *Let X and Y be Borel spaces, with X countably separated, and let f be a Borel map of X onto Y. Assume that f has a Borel cross section $g: Y \to X$. Then $g(Y)$ is a Borel set in X and g is a Borel isomorphism of Y onto $g(Y)$. In particular, if X is a standard Borel space then so is Y.*

PROOF. Let $F \subseteq X$ be the set of all fixed points of the Borel function $g \circ f: X \to X$. One sees easily that $F = g(Y)$, and we claim that F is a Borel set in X. Indeed, because X is countably separated, the diagonal $\Delta = \{(x, x): x \in X\}$ is a Borel set in $X \times X$ (see the proof of 3.3.1). Therefore since $x \to (x, g \circ f(x))$ is a Borel map of X into $X \times X$ we conclude that $F = \{x \in X: (x, g \circ f(x)) \in \Delta\}$ is a Borel set.

Finally, $g: Y \to F$ and $f|_F: F \to Y$ are Borel maps inverse to one another, so in particular g is an isomorphism of Y onto F. □

Note that the set $g(Y)$ in 3.4.2 is a *Borel transversal* for the equivalence relation $x \sim y$ iff $f(x) = f(y)$; i.e., it is a Borel set in X which meets each equivalence class in a single point.

These are the only cross section results required for our present purposes, and the reader may omit the rest of this section without essential loss.

We want to discuss "usable" cross sections in general, and to draw a significant conclusion about why they may fail to exist. In the preceding section we saw that if a Borel space Y is the range of, say, a standard Borel space X under a Borel map $f: X \to Y$, then Y is analytic if and only if it is countably separated. Thus one might expect to have a Borel cross section for f when Y is countably separated. However, if Y is merely analytic but not standard, then 3.4.2 shows that f has no Borel cross section. One might then conjecture that Borel cross sections exist when both Y and X are standard. Unfortunately, even that is not true; D. Blackwell has given an example of a Borel subset X of the unit square $[0, 1] \times [0, 1]$ such that if $f(x, y) = x$ for $(x, y) \in X$, then $f(X) = [0, 1]$ (a Polish space), while f does not have a Borel cross section [3].

In spite of all this, one can do almost as well in the case where Y is countably separated. The concept we need has to do with measurability. Let $(X, \mathscr{B})$ be a Borel space. A subset $A \subseteq X$ is called *absolutely measurable* if A is μ-measurable for every finite Borel measure μ on X; equivalently, for every finite Borel measure μ there exist Borel sets E_μ, F_μ such that $E_\mu \subseteq A \subseteq F_\mu$ and $\mu(F_\mu \backslash E_\mu) = 0$. The family $\mathscr{A}$ of all absolutely measurable sets is a σ-field containing $\mathscr{B}$, indeed it is the intersection $\bigcap_\mu \mathscr{A}_\mu$, where $\mathscr{A}_\mu$ denotes the σ-field of all μ-measurable subsets of X, and μ runs over all finite Borel measures. Because analytic sets in Polish spaces are absolutely measurable (3.2.4) it follows that $\mathscr{A}$ is generally larger than $\mathscr{B}$. A function g from one Borel space into another is called *absolutely measurable* if the inverse image under g of every Borel set is absolutely measurable; equivalently, g is μ-measurable for every finite Borel measure μ on the domain of g. Thus, absolutely measurable functions are amenable to measure theoretic analysis, and from that point of view are just as good as Borel functions. We now prove that in all "reasonable" cases, absolutely measurable cross sections do exist.

Theorem 3.4.3. *Let X be an analytic Borel space, let Y be a countably separated Borel space, and let f be a Borel map of X onto Y. Then f has an absolutely measurable cross section.*

Proof. By Corollary 1 of 3.3.5 Y is an analytic Borel space, so there is no loss if we assume that X and Y are analytic subspaces of Polish spaces P and Q. The map $x \mapsto (x, f(x)) \in P \times Q$ is a Borel map of X onto the graph of f, so by 3.3.4 graph f is an analytic subset of $P \times Q$. Thus graph f is the continuous image of a Polish space and by 3.1.4 we may take the latter to be N^∞; so let $\varphi: N^\infty \to P \times Q$ be continuous, $\varphi(N^\infty) =$ graph f. Finally, let p_2 be the projection of $P \times Q$ onto Q. Then $p_2 \circ \varphi: N^\infty \to Q$ is continuous and has range Y. We will find an absolutely measurable cross section $g: Y \to N^\infty$ for $p_2 \circ \varphi$. Granting that for a moment, let p_1 be the projection of $P \times Q$ onto P. Then note that the function $p_1 \circ \varphi \circ g$ maps Y into X, it is absolutely measurable (because $p_1 \circ \varphi$ is Borel and g is absolutely measurable), and it is a cross section for f because $f \circ p_1 = p_2$ on graph f and $p_2 \circ \varphi \circ g = \mathrm{id}_Y$ implies $f \circ (p_1 \circ \varphi \circ g) = \mathrm{id}_Y$. Thus the theorem will follow if we can find such a function g.

Now by continuity $(p_2 \circ \varphi)^{-1}(\{y\})$ is closed in N^∞, for each $y \in Y$, and as in 3.4.1 we define $g(y)$ to be the smallest element in that set. Then g is a cross section for $p_2 \circ \varphi$ and, exactly as in 3.4.1, $g^{-1}(-\infty, v) = p_2 \circ \varphi(-\infty, v)$ for every $v \in N^\infty$, and we need only prove that $p_2 \circ \varphi(-\infty, v)$ is an absolutely measurable set in Y. Since $(-\infty, v)$ is open and therefore a Polish subspace, $p_2 \circ \varphi(-\infty, v)$ is an analytic set in Q and thus measurable for every finite measure on Q (3.2.4). So if μ is any finite Borel measure on Y, then $p_2 \circ \varphi(-\infty, v)$ is measurable with respect to the Borel measure $\sigma(E) = \mu(E \cap Y)$ on Q, and therefore it is μ-measurable as well. □

We conclude that if $\sim$ is an equivalence relation in an analytic Borel space X for which $X/\sim$ is countably separated, then the canonical projection $q: X \to X/\sim$ has an absolutely measurable cross section.

We now take up a common situation in which measurable cross sections do *not* exist. Some terminology will be convenient. Let $\sim$ be an equivalence relation in a Borel space X; though it is not essential we will assume for simplicity that each equivalence class $c(x) = \{y \in X: y \sim x\}$ is a Borel set. Let $\mathscr{S}$ be the sub σ-field of all $\sim$-saturated Borel subsets of X (see the discussion preceding Example 3.3.3). A Borel probability measure μ on X is called *ergodic* (with respect to $\sim$) if $\mu(E) = 0$ or 1 for every saturated Borel set E. The terminology derives from the case where $\sim$ is the equivalence relation induced in X by a group of transformations. Thus μ is ergodic iff the restriction of μ to $\mathscr{S}$ is $\{0, 1\}$-valued. One example of this is any measure μ concentrated on a single equivalence class ($\mu(c(x)) = 1$ for some $x \in X$). Such an ergodic measure is called *trivial*. The following lemma implies that, in the familiar Borel spaces, the trivial equivalence relation $x \sim y$ iff $x = y$ has only trivial ergodic measures.

Lemma. *Let X be a countably generated Borel space, and let μ be a $\{0, 1\}$-valued Borel probability measure on X. Then μ is a point mass.*

PROOF. By the lemma preceding 3.3.5 we may assume X is a subspace of a Polish space P. Let μ be a $\{0, 1\}$-valued measure on X, and define a Borel measure ν on P by $\nu(E) = \mu(X \cap E)$. It suffices to show that there is a point $p \in P$ for which $\nu(\{p\}) = 1$, and that will follow if we show that the closed support of ν (i.e., the complement of the union of all open sets $V \subseteq P$ for which $\nu(V) = 0$) is a singleton (here we use the fact that nonzero Borel measures in Polish spaces have nonvoid closed supports). Assume not, say $p \neq q$ are in the closed support of ν. Then there are disjoint open neighborhoods U, V of p and q, and hence $\nu(U) > 0$ and $\nu(V) > 0$. Since $\nu(U) + \nu(V) = \nu(U \cup V) \leqslant 1$, this contradicts the fact that ν (like μ) is $\{0, 1\}$-valued. □

Proposition. *Let X be a countably generated Borel space, let $\sim$ be an equivalence relation in X, and let $q: X \to X/\sim$ be the canonical projection. Assume that for every finite Borel measure ν on $X/\sim$ there is a ν-measurable cross section h_ν for q. Then every ergodic Borel measure on X is trivial.*

PROOF. Choose an ergodic Borel probability measure μ on X. Then we may define a Borel measure ν on $X/\sim$ by $\nu(E) = \mu(q^{-1}(E))$. Since μ is ergodic ν is $\{0, 1\}$-valued, and thus so is the completion $\bar{\nu}$ of ν. By hypothesis there is a ν-measurable cross section h_ν for q, and thus we may define a third Borel measure μ_0 on X via $\mu_0(F) = \bar{\nu}(h_\nu^{-1}(F))$. Now μ_0 is $\{0, 1\}$-valued so by the lemma there is a point $p \in X$ for which $\mu_0(\{p\}) = 1$. Let $y = q(p)$. Note that $\{y\} = h_\nu^{-1}(\{p\})$ (because h_ν is a cross section for q) and $\{y\}$ is a Borel set in $X/\sim$ because $c(p) = q^{-1}(\{y\})$ is a Borel set in X. Thus $\nu(\{y\}) = \bar{\nu}(\{y\}) = \mu_0(\{p\}) = 1$, and it follows that $\mu(c(p)) = 1$. Thus μ is trivial. □

In particular, we conclude that *if an equivalence relation $\sim$ in a Polish space P admits a nontrivial ergodic measure, then the projection $q: P \to P/\sim$ does not have an absolutely measurable cross section.* Here is one example of this. Let P be the unit circle $\{|z| = 1\}$ with its usual Borel structure, let t be any real number not a rational multiple of π, and define $z_1 \sim z_2$ in P if there is an integer n such that $z_2 = e^{int}z_1$. Then each equivalence class is a countable subset of P (therefore an F_σ). So if we let μ be normalized Lebesgue measure on P then every equivalence class has measure zero. However, μ is well-known to be an ergodic measure for $\sim$ (the easiest way to see that a saturated Borel set E satisfies $\mu(E) = 0$ or 1 is to look at the Fourier coefficients of the function χ_E, noting that $\chi_E(e^{it}z) = \chi_E(z)$ for all $z \in P$). Therefore the projection q of P onto the orbit space $P/\sim$ has no absolutely measurable section.

Of course the axiom of choice implies that cross sections exist for q, but in the precise sense above they are all too pathological to be of any use in analysis (note, incidentally, the close parallel between such functions and Kolmogorov's example of a non Lebesgue measurable set). This suggests that quotient spaces as in the example are inaccessible in an essential way: their properties may be undecidable within the usual processes of analysis. The precise formulation and proof of that conjecture is, however, a problem in metamathematics, not analysis.

In Section 4.1 we will discuss the spectrum of an arbitrary (separable) C^*-algebra. If the C^*-algebra is *not* GCR, it is known that its spectrum is a quotient space of the above pathological type. So the preceding discussion strongly indicates that there is no hope of giving a concrete parametric description of the irreducible representations of such a C^*-algebra.

From Commutative Algebras to GCR Algebras 4

In this chapter we will work out the representation theory of separable GCR algebras. The basic results, appearing at the end of Section 4.3, are presented so as to obviously generalize the results of Section 2.2 which classify normal operators and representations of abelian C^*-algebras. But in contrast with the commutative case, it is necessary here to make essential use of the material on Borel structures from Chapter 3.

The spectrum of an abstract C^*-algebra is defined in Section 4.1, and we develop its properties as a Borel space. The main result asserts that the spectrum of a separable GCR algebra is a standard Borel space.

Section 4.2 contains a discussion of some results relating to reduction theory that are needed in the last section. This material has been kept to the barest minimum. Nevertheless, as the expert will notice, it is still possible to deduce, say, the central decomposition for separably-acting von Neumann algebras from these results. We have resisted the temptation to do that here.

4.1. The Spectrum of a C^*-algebra

We have seen two ways by which the spectrum of a commutative C^*-algebra contributes to the analysis of the algebra. On the one hand it provides a concrete representation for elements in the algebra as complex-valued continuous functions, and on the other it forms a conveniently topologized "parameter" space for the irreducible representations of the algebra (with each point p we associate the one-dimensional representation "evaluation at p"). We are now going to generalize the notion of spectrum to non-commutative C^*-algebras.

The first thing to notice is that there are a number of possible candidates. Even in the commutative case one may view the spectrum as a space of maximal ideals (endowed with the hull-kernel topology) or as a space of complex homomorphisms (with the relative weak*-topology). In the noncommutative case the proper counterparts of maximal ideals are not maximal ideals but primitive ideals (a closed two sided ideal K in a C^*-algebra A is *primitive* if there is a nonzero irreducible representation π of A on a Hilbert space such that $K = \text{kernel } \pi$), and the counterparts of complex homomorphisms are irreducible representations. In place of the maximal ideal space one has the set prim A of all primitive ideals of A endowed with the hull-kernel topology: by definition a subset $S \subseteq \text{prim } A$ is closed iff every primitive ideal which contains the intersection $\bigcap K$, $K \in S$, is already in S. The topological space prim A is called the *structure space* of A, and was introduced in general ring theory, in a somewhat different way, by N. Jacobson ([6], Section 3). We shall not make use of the structure space, but we do want to point out one or two of its properties.

Let us first consider an example. Let A be the family of all continuous functions F defined on the unit interval $[0, 1]$, taking values in the C^*-algebra M_2 of all complex 2×2 matrices, such that $F(0)$ is diagonal. Such an F can be regarded as a 2×2 matrix (f_{ij}) of continuous complex valued functions on $[0, 1]$ satisfying $f_{12}(0) = f_{21}(0) = 0$. A becomes a C^*-algebra with respect to the pointwise operations and the norm $\|F\| = \sup_t \|F(t)\|$. If we cause M_2 to act in the obvious way on a two dimensional Hilbert space $\mathscr{H}$, then for each t, $0 < t \leqslant 1$, we obtain an irreducible representation π_t of A on $\mathscr{H}$ by evaluation at $t: \pi_t(F) = F(t)$. Associated with the point $t = 0$ we have two one-dimensional irreducible representations, namely $\alpha(F) = f_{11}(0)$ and $\beta(F) = f_{22}(0)$. It can be shown that these are the only irreducible representations of A (to within unitary equivalence) and therefore prim A is identified with the union $\{\alpha, \beta\} \cup (0, 1]$ of the interval $(0, 1]$ with *two* additional points α, β in place of 0. The hull-kernel topology restricts to the usual topology on $(0, 1]$, whereas basic neighborhoods of α (resp. β) look like $\{\alpha\} \cup (0, \varepsilon)$ (resp. $\{\beta\} \cup (0, \varepsilon)$), for $0 < \varepsilon < 1$. Thus every neighborhood of α must intersect every neighborhood of β, and so prim A is *not* a Hausdorff space. Note, incidentally, that A is even a CCR algebra.

In the general case prim A is not even a T_1 space (the algebraic reason being that one primitive ideal can be contained properly in another), though it is always at least T_0 ([6], Section 3). But there are even worse difficulties. Suppose for example that A is a simple C^*-algebra, i.e., A has no closed ideals other than 0 and A. In particular there are no primitive ideals other than 0, and therefore prim A consists of a single point. Since there exist simple C^*-algebras having many inequivalent irreducible representations (all having kernel 0), it is apparent that prim A will not be adequate to serve as a parameter space for the irreducible representations of A. For this reason the spectrum is defined in terms of irreducible representations rather than ideals.

Roughly speaking, the spectrum $\hat{A}$ of a C^*-algebra A is the set of all equivalence classes of irreducible representations of A, endowed with the "natural" Borel structure. When A is commutative the set $\hat{A}$ will turn out to be set of all complex homomorphisms of A with the Borel structure generated by the weak*-closed subsets. It is also possible in general to define a topology on $\hat{A}$. But as the preceding paragraphs suggest, this topology generally has such poor separation properties that it loses much (though by no means all) of its usefulness. Moreover, since the construction and classification of representations involves measures rather than continuous functions, it is the Borel structure rather than the topology that we shall have to deal with. We remark that this view of the spectrum of a C^*-algebra is due to Mackey [20].

Let us now make all of this more precise. Let A be a separable C^*-algebra, fixed throughout the remainder of this section. Let $\mathscr{H}$ be a separable Hilbert space, and consider the set rep$(A, \mathscr{H})$ of all representations of A on $\mathscr{H}$; the elements of rep$(A, \mathscr{H})$ are allowed to be degenerate. For each $x \in A$, $\xi, \eta \in \mathscr{H}$, we have a complex-valued function $\pi \to (\pi(x)\xi, \eta)$ defined on rep$(A, \mathscr{H})$, and we may topologize rep$(A, \mathscr{H})$ by taking the weakest topology making all of these functions continuous. For $x \in A$, ξ, $\eta \in \mathscr{H}$, we have $\|\pi(x)\xi - \eta\|^2 = (\pi(x^*x)\xi, \xi) - 2\mathscr{Re}(\pi(x)\xi, \eta) + \|\eta\|^2$, so that $\pi \to \pi(x)\xi$ defines a continuous function from rep$(A, \mathscr{H})$ into $\mathscr{H}$ for each x, ξ, and evidently we could have used these functions to define the topology on rep$(A, \mathscr{H})$. Now choose a dense sequence $x_1, x_2, \ldots$ (resp. $\xi_1, \xi_2, \ldots$) in the unit ball of A (resp. $\mathscr{H}$) and put

$$d(\pi, \pi') = \sum_{m, n=1} 2^{-m-n}\|\pi(x_m)\xi_n - \pi'(x_m)\xi_n\|.$$

Then d defines a metric on rep$(A, \mathscr{H})$ which makes rep$(A, \mathscr{H})$ into a metric space. It is not hard to see that rep$(A, \mathscr{H})$ is complete and this topology has a countable base (we omit these details; see [9]), so that rep$(A, \mathscr{H})$ becomes a Polish space. The set irr$(A, \mathscr{H})$ of all irreducible members of rep$(A, \mathscr{H})$ thus becomes a topological space in its relative topology.

Now for each $n = \infty, 1, 2, \ldots$ (where as usual ∞ stands for $\aleph_0$) choose a fixed Hilbert space $\mathscr{H}_n$ of dimension n. For each n we have a natural equivalence relation $\sim$ in irr$(A, \mathscr{H}_n)$, namely $\pi \sim \sigma$ iff π and σ are unitarily equivalent. Give irr$(A, \mathscr{H}_n)$ the Borel structure generated by its topology, and then define the "n-dimensional" part $\hat{A}_n$ of the spectrum of A as the Borel quotient space irr$(A, \mathscr{H}_n)/\sim$. Finally, the *spectrum* $\hat{A}$ of A is defined as the direct sum of the Borel spaces $\hat{A}_\infty, \hat{A}_1, \hat{A}_2, \ldots$.

Note first that every nonzero irreducible representation of A is represented in $\hat{A}$. More precisely, if π is an irreducible representation of A on some Hilbert space $\mathscr{K}$, then for each nonzero vector $\xi \in \mathscr{K}$ $\pi(A)\xi$ is dense in $\mathscr{K}$, and since A has a countable dense subset it follows that $\mathscr{K}$ is separable. So if n is the dimension of $\mathscr{H}$ then π is equivalent to some element of irr$(A, \mathscr{H}_n)$, as asserted.

We shall now collect a few general properties of $\hat{A}$.

Theorem 4.1.1. *Let A be a C^*-algebra, let $\mathscr{H}$ be a separable Hilbert space, and let T be a bounded self-adjoint operator on $\mathscr{H}$. Then the set $\mathscr{S}_T$ of all* *representations* $\pi \in \operatorname{rep}(A, \mathscr{H})$ *for which* T *belongs to the weak closure of* $\pi(A)$ *is a* G_δ *in* $\operatorname{rep}(A, \mathscr{H})$.

PROOF. Choose any real number $r > \|T\|$ and choose $\pi \in \operatorname{rep}(A, \mathscr{H})$. We claim: $\pi \in \mathscr{S}_T$ iff T belongs to the weak closure of $\pi(A_r)$, A_r denoting the ball of radius r in A. Indeed the "if" part is trivial, so assume $\pi \in \mathscr{S}_T$. By the double commutant theorem T belongs to the weak closure of the $*$-algebra $\pi(A)$, so by Kaplansky's density theorem (1.2.2) there is a net $x_n \in A$ such that $\|\pi(x_n)\| \leqslant \|T\|$ for all n and $\lim_n \pi(x_n) = T$ in the weak operator topology. Now if K is the kernel of π then by 1.3.2, $\|\pi(x)\| = \inf\|x + k\|$, $k \in K$, so we may choose the x_n above in such a way that $\|x_n\|$ is as close to $\|T\|$ as we please. In particular we can choose $\|x_n\| \leqslant r$, and the claim follows.

Thus $\mathscr{S}_T$ may be written as $\{\pi \in \operatorname{rep}(A, \mathscr{H}); T \in \pi(A_r)^{-w}\}$, and it suffices to show that the latter is a G_δ.

Now choose a countable dense subset $\xi_1, \xi_2, \ldots$ in $\mathscr{H}$. For each $n \geqslant 1$, $x \in A$, define a real function $f_{n,x}$ on $\operatorname{rep}(A, \mathscr{H})$ by

$$f_{n,x}(\pi) = \sum_{i,j=1}^{n} |(T\xi_i - \pi(x)\xi_i, \xi_j)|.$$

Now $\|\pi(A_r)\| \leqslant r$ for each π, so that $T \in \pi(A_r)^{-w}$ if and only if for every n, $k \geqslant 1$ there is an $x \in A_r$ such that $f_{n,x}(\pi) < 1/k$. Symbolically,

$$\mathscr{S}_T = \bigcap_{n=1}^{\infty} \bigcap_{k=1}^{\infty} \bigcup_{x \in A_r} \{\pi \in \operatorname{rep}(A, \mathscr{H}) : f_{n,x}(\pi) < 1/k\},$$

and since each function $f_{n,x}$ is continuous, this formula exhibits $\mathscr{S}_T$ as a G_δ. □

Note that the postulated separability of A did not enter the above proof, and in fact 4.1.1. and its corollaries are valid for separable representations of *inseparable* C^*-algebras.

Corollary 1. *The set of nondegenerate elements of* $\operatorname{rep}(A, \mathscr{H})$ *is a* G_δ.

PROOF. This is $\mathscr{S}_I$ where I is the identity operator on $\mathscr{H}$. □

Corollary 2. $\operatorname{irr}(A, \mathscr{H})$ *is a* G_δ *in* $\operatorname{rep}(A, \mathscr{H})$.

PROOF. Let S and T be the real and imaginary parts of any particular irreducible operator on $\mathscr{H}$. Then a C^*-algebra on $\mathscr{H}$ is irreducible iff its weak closure contains both S and T. Thus $\operatorname{irr}(A, \mathscr{H}) = \mathscr{S}_S \cap \mathscr{S}_T$ is an intersection of two G_δ's in $\operatorname{rep}(A, \mathscr{H})$, and the conclusion follows. □

Corollary 3. *For every* $\pi_0 \in \operatorname{rep}(A, \mathscr{H})$, $\{\sigma \in \operatorname{rep}(A, \mathscr{H}) : \sigma \propto \pi_0\}$ *is a* G_δ *in* $\operatorname{rep}(A, \mathscr{H})$.

PROOF. Let $I \oplus 0$ be the projection of $\mathscr{H} \oplus \mathscr{H}$ onto its first coordinate space. It follows from 2.1.4 that $\sigma \propto \pi_0$ iff $I \oplus 0$ belongs to the weak closure of $\sigma \oplus \pi_0(A)$ (in $\mathscr{L}(\mathscr{H} \oplus \mathscr{H})$). Now by 4.1.1, $\{\lambda \in \text{rep}(A, \mathscr{H} \oplus \mathscr{H}) : I \oplus 0 \in \lambda(A)^-\}$ is a G_δ in rep$(A, \mathscr{H} \oplus \mathscr{H})$; and since $\sigma \to \sigma \oplus \pi_0$ defines a continuous map of rep$(A, \mathscr{H})$ into rep$(A, \mathscr{H} \oplus \mathscr{H})$, we conclude that $\{\sigma \in \text{rep}(A, \mathscr{H}) : \sigma\ \pi_0\}$, being the inverse image of a G_δ under a continuous map, is a G_δ. □

Corollary 4. *For every $\pi_0 \in$ irr$(A, \mathscr{H})$, the equivalence class $\{\pi \in$ irr$(A, \mathscr{H})$: $\pi \sim \pi_0\}$ is an F_σ subset of the topological space* irr$(A, \mathscr{H})$.

PROOF. Let C be the complement of $\{\pi : \pi \sim \pi_0\}$ in irr$(A, \mathscr{H})$. It suffices to show that C is a G_δ in irr$(A, \mathscr{H})$. Now two *irreducible* representations of A on $\mathscr{H}$ are inequivalent if and only if they are disjoint. Thus we have $C =$ irr$(A, \mathscr{H}) \cap \{\pi \in \text{rep}(A, \mathscr{H}) : \pi \mathrel{\delta} \pi_0\}$, and the conclusion follows from Corollary 3. □

Let us now examine $\hat{A}_n$ for a separable C^*-algebra A. Corollary 2 and the remarks preceding 4.1.1 show that irr$(A, \mathscr{H}_n)$ is a G_δ in the Polish space rep$(A, \mathscr{H}_n)$. By 3.1.2 irr$(A, \mathscr{H}_n)$ is therefore a *Polish space* in its relative topology; and in particular the Borel structure on irr$(A, \mathscr{H}_n)$ is standard. The inverse image of each point in $\hat{A}_n$ is a single equivalence class of elements in irr$(A, \mathscr{H}_n)$, which by Corollary 4 is a Borel (in fact F_σ) set in irr$(A, \mathscr{H}_n)$. These statements remain valid of course for the disjoint union $\hat{A} = \hat{A}_\infty \cup \hat{A}_1 \cup \hat{A}_2 \cup \cdots$, and this proves the following description, due to Mackey [20], of the dual of an arbitrary separable C^*-algebra A.

Theorem 4.1.2. *$\hat{A}$ is a quotient space of a standard Borel space, in which each singleton is a Borel set.*

We have already indicated in a vague way that $\hat{A}$ can be made into a topological space, but one whose topology is bad in that points are not necessarily closed. 4.1.2 asserts that points are at least "definable."

It is natural to ask at this point if the Borel space $\hat{A}$ is a good one or a pathological one. Since $\hat{A}$ is a quotient of a standard Borel space, the results of Chapter 3 show that the best one would hope for in general is that $\hat{A}$ be analytic, and Corollary 2 of 3.3.5 shows that this is equivalent to having $\hat{A}$ countably separated. It turns out that if A is a GCR algebra then $\hat{A}$ is in fact *standard*, and if A is not GCR then $\hat{A}$ is not even countably separated. The second assertion follows from a deep theorem of Glimm, and the interested reader should consult [13] for the details. The first one, however, we will deduce as a corollary to the following theorem on cross sections.

Theorem 4.1.3. *Let A be a separable* GCR *algebra, fix $n = \infty, 1, 2, \ldots$, a denumerable cardinal, and let q:*irr$(A, \mathscr{H}_n) \to \hat{A}_n$ *be the canonical quotient map. Then q has a Borel cross section.*

PROOF. Since irr$(A, \mathscr{H}_n)$ is a Polish space and since the equivalence relation in irr$(A, \mathscr{H}_n)$ is defined by a group of homeomorphisms (the unitary group

$\mathscr{U}$ of $\mathscr{H}_n$ acts on irr$(A, \mathscr{H}_n)$ in the obvious way: for $U \in \mathscr{U}$ and $\pi \in$ irr$(A, \mathscr{H})$, $U \cdot \pi$ is the representation $x \mapsto U\pi(x)U^*$), the corollary of 3.4.1 would yield the required cross section provided that for each $\pi \in$ irr$(A, \mathscr{H}_n)$ the equivalence class $C_\pi = \{\sigma \in \text{irr}(A, \mathscr{H}_n): \sigma \sim \pi\}$ is closed. Unfortunately this is not true, even when A is a GCR algebra (though by Corollary 4 above C_π is in general at worst a countable union of closed sets). What we will do instead is produce a countable partition $\{E_1, E_2, \ldots\}$ of irr$(A, \mathscr{H}_n)$ into disjoint G_δ's such that each E_j is saturated with respect to equivalence (i.e., $\pi \in E_j$, $\sigma \sim \pi$ implies $\sigma \in E_j$) and such that C_π *is closed in the relative topology of* E_j, for every $\pi \in E_j$. Since each E_j is a Polish space (3.1.2) we may apply the corollary of 3.4.1 to obtain a Borel cross section h_j for $q|_{E_j}$. Because of the disjointness and saturation of $\{E_j\}$ it will therefore follow that the union of the functions $h_1, h_2, \ldots$ is the desired Borel cross section for q.

Let us now construct $\{E_j\}$. Since A is GCR it has a composition series $\{J_\rho: 0 \leqslant \rho \leqslant \alpha\}$ of ideals as in 1.5.5, where the ordinal α is countable by separability of A (cf. remark following 1.5.5). For each ordinal ρ, $0 \leqslant \rho \leqslant \alpha$, let $F_\rho = \{\pi \in \text{irr}(A, \mathscr{H}_n): \pi(J_\rho) \neq \{0\}\}$. It is clear that F_ρ is saturated with respect to equivalence, and a routine check (which we leave for the reader) shows that F_ρ is open in irr$(A, \mathscr{H}_n)$. Therefore each set $F_{\rho+1}\backslash F_\rho$, $0 \leqslant \rho < \alpha$, being the intersection of a saturated open set with a saturated closed set, is a saturated G_δ. The family $\{F_{\rho+1}\backslash F_\rho: 0 \leqslant \rho < \alpha\}$ is countable (since α is a countable ordinal), and its elements are mutually disjoint. Thus it suffices to show that $\bigcup_{0 \leqslant \rho < \alpha} F_{\rho+1}\backslash F_\rho = \text{irr}(A, \mathscr{H}_n)$, and that for each $\rho < \alpha$ and $\pi \in F_{\rho+1}\backslash F_\rho$, C_π is closed in $F_{\rho+1}\backslash F_\rho$.

For the first conclusion, choose $\pi \in$ irr$(A, \mathscr{H}_n)$. Since $\pi(J_\alpha) = \pi(A) \neq 0$ and $\pi(J_0) = \pi(0) = 0$, there is a first ordinal $\lambda > 0$ such that $\pi(J_\lambda) \neq 0$. Note that λ cannot be a limit ordinal: for if it were then J_λ would be the norm closure of $\bigcup_{\mu < \lambda} J_\mu$, and thus $\pi(J_\mu) = 0$ for $\mu < \lambda$ would imply $\pi(J_\lambda) = 0$, contradicting the choice of λ. Thus λ has an immediate predecessor $\lambda - 1$, and so $\pi \in F_\lambda\backslash F_{\lambda-1} \subseteq \bigcup_{0 \leqslant \rho < \alpha} F_{\rho+1}\backslash F_\rho$. This proves that irr$(A, \mathscr{H}_n) = \bigcup_{0 \leqslant \rho < \alpha} F_{\rho+1}\backslash F_\rho$.

It remains to prove that each equivalence class is closed in $F_{\rho+1}\backslash F_\rho$. Therefore choose $\pi \in F_{\rho+1}\backslash F_\rho$, and let σ be an element of $(F_{\rho+1}\backslash F_\rho) \cap C_\pi^-$. We claim that σ is equivalent to π (this will complete the proof). Since $\sigma(J_{\rho+1}) \neq 0$ and $\pi(J_{\rho+1}) \neq 0$ (both σ and π are in $F_{\rho+1}$) and since $J_{\rho+1}$ is an ideal, it suffices to show that the restrictions of σ and π to $J_{\rho+1}$ are equivalent (1.3.4). Since both σ and π annihilate J_ρ (neither belongs to F_ρ) we may regard these restrictions as representations σ', π', respectively, of the CCR quotient $J_{\rho+1}/J_\rho$. Note that σ' and π' are still irredicible, by (1.3.4), and we are now reduced to proving that σ' and π' are equivalent. Now since σ belongs to the closure of C_π, it is apparent that σ' belongs to the closure of $C_{\pi'} = \{\lambda \in \text{irr}(J_{\rho+1}\backslash J_\rho, \mathscr{H}_n): \lambda \sim \pi'\}$ (just choose a sequence U_n of unitary operators such that $U_n \pi U_n^*$ converges to σ, then restrict everything to $J_{\rho+1}$ and pass to the quotient $J_{\rho+1}/J_\rho$). Finally, we claim that $\ker \pi' \subseteq \ker \sigma'$. Indeed if $x \in J_{\rho+1}/J_\rho$ is such that $\pi'(x) = 0$, then $\lambda'(x) = 0$ for every

$\lambda' \in C_{\pi'}$, and since σ' is a limit of such λ's, we see that $\lambda'(x) = 0$. Since $J_{\rho+1}/J_\rho$ is a CCR algebra, we can conclude from 1.5.2 that π' is equivalent to σ', and the proof is now complete. □

Corollary. *If A is a separable* GCR *algebra, then $\hat{A}$ is a standard Borel space.*

PROOF. For each $n = \infty, 1, 2, \ldots$, the canonical quotient map of $\operatorname{irr}(A, \mathscr{H}_n)$ onto $\hat{A}_n$ has a Borel cross section; by 3.4.2 $\hat{A}_n$ is standard, and therefore $\hat{A}$, being a countable direct sum of standard Borel spaces, is standard. □

Theorem 4.1.3 was first obtained by Dixmier [9], in a slightly different form. Its corollary is due to Dixmier [10], Fell [12], and Glimm [13].

Theorem 4.1.3 above asserts that Borel cross sections exist (at least for separable GCR algebras), and the question arises as to what extent they are unique. We conclude this section with a discussion of uniqueness. The result is a consequence of the following lemma, which is of some interest in its own right. First, we need some preliminaries. Let $\mathscr{U}$ be the group of unitary operators on a separable Hilbert space $\mathscr{H}$. It is a familiar fact that the strong and weak operator topologies coincide on $\mathscr{U}$, therefore $\mathscr{U}$ is unambiguously a topological space in either of these topologies, with respect to which it is a subspace of the Polish space $\{T \in \mathscr{L}(\mathscr{H}): \|T\| \leqslant 1\}$ (the weak operator topology on the latter). We want to point out that $\mathscr{U}$ is a Polish space. For that choose a countable dense set of unit vectors $w_1, w_2, \ldots$ in $\mathscr{H}$ and define a metric d on $\mathscr{U}$ by

$$d(U, V) = \sum_{n=1}^{\infty} 2^{-n}(\|Uw_n - Vw_n\| + \|U^*w_n - V^*w_n\|).$$

A simple argument which we omit shows that d is a *complete* metric on $\mathscr{U}$, and thus $\mathscr{U}$ is Polish.

Now $\mathscr{U}$ is also a group, and its center is the compact subgroup $\mathbb{T}$ of all scalar operators λI, $|\lambda| = 1$. Consider the quotient group $\mathscr{U}/\mathbb{T}$, with the Borel structure generated by its quotient topology. It turns out that $\mathscr{U}/\mathbb{T}$ is Polish as a topological space, but all we shall require is that $\mathscr{U}/\mathbb{T}$ is a standard Borel space. One way to see this is to note that the quotient map $p: \mathscr{U} \to \mathscr{U}/\mathbb{T}$ satisfies the hypotheses of Theorem 3.4.1 and therefore has a Borel cross section. By Proposition 3.4.2 $\mathscr{U}/\mathbb{T}$ is Borel isomorphic to a Borel subset of the Polish space $\mathscr{U}$, which shows $\mathscr{U}/\mathbb{T}$ is standard.

Lemma 4.1.4. *Let A be a separable C^*-algebra, let $\mathscr{H}$ be a separable Hilbert space, and let Γ be the graph of the unitary equivalence relation in* $\operatorname{irr}(A, \mathscr{H})$*, regarded as a subset of the Polish space* $\operatorname{irr}(A, \mathscr{H}) \times \operatorname{irr}(A, \mathscr{H})$.

Then Γ is a Borel set, and there is a Borel map $\gamma \mapsto V_\gamma$ of Γ into the unitary group $\mathscr{U}$ of $\mathscr{H}$ such that $V_{(\pi,\sigma)}\pi V^*_{(\pi,\sigma)} = \sigma$ for all $(\pi, \sigma) \in \Gamma$.

PROOF. Define a map ϕ of the Polish space $\operatorname{irr}(A, \mathscr{H}) \times \mathscr{U}$ into the Polish space $\operatorname{irr}(A, \mathscr{H}) \times \operatorname{irr}(A, \mathscr{H})$ by $\phi(\pi, V) = (\pi, V\pi V^*)$. Clearly ϕ is continuous, maps onto Γ, and the inverse image of each point in $\text{irr}(A, \mathscr{H}) \times \text{irr}(A, \mathscr{H})$ is closed. We want to show now that ϕ maps open sets to Borel sets. Note first that $\phi(\pi, V) = \phi(\pi', V')$ iff $\pi = \pi'$ and $V = V' \pmod{\mathbb{T}}$. Thus ϕ lifts to a continuous map $\dot{\phi}$ of the topological space $\text{irr}(A, \mathscr{H}) \times \mathscr{U}/\mathbb{T}$ onto Γ in the usual way. Now $\dot{\phi}$ is clearly 1—1 and it is a Borel map (by continuity) if we give $\text{irr}(A, \mathscr{H}) \times \mathscr{U}/\mathbb{T}$ the product Borel structure. By the remarks preceding the lemma, the domain of $\dot{\phi}$ is a standard Borel space. Thus by 3.3.2, Γ is a Borel set and $\dot{\phi}$ is a Borel isomorphism. In particular $\dot{\phi}$ maps open sets to Borel sets. Since the quotient map $p:\mathscr{U} \to \mathscr{U}/\mathbb{T}$ is open, the map $\text{id} \times p:(\pi, V) \in \text{irr}(A, \mathscr{H}) \times \mathscr{U} \mapsto (\pi, p(V)) \in \text{irr}(A, \mathscr{H}) \times \mathscr{U}/\mathbb{T}$ is open, and it is therefore apparent that $\phi = \dot{\phi} \circ (\text{id} \times p)$ maps open sets to Borel sets.

By Theorem 3.4.1, there is a Borel map $h:\Gamma \to \text{irr}(A, \mathscr{H}) \times \mathscr{U}$ such that $\phi \circ h$ is the identity on Γ. Writing $h(\gamma)$ in coordinates (π_γ, V_γ), it follows from an inspection of the formula $\phi \circ h = \text{id}$ that the Borel map $\gamma - V_\gamma$ has the stated properties. □

Corollary Uniqueness of cross sections. *Let A be a separable C^*-algebra, let X be a Borel space, and let $\mathscr{H}$ be a separable Hilbert space. Let $x \mapsto \pi_x$, $x \mapsto \sigma_x$ be any two Borel maps of X into* $\text{irr}(A, \mathscr{H})$ *such that $\pi_x \sim \sigma_x$ for all $x \in X$. Then there is a Borel map $x \mapsto U_x$ of X into the unitary group $\mathscr{U}$ of $\mathscr{H}$ such that $\sigma_x = U_x \pi_x U_x^*$ for all $x \in X$.*

Proof. Let $\Gamma \subseteq \text{irr}(A, \mathscr{H}) \times \text{irr}(A, \mathscr{H})$ be the graph of the relation $\sim$. By the lemma there is a Borel map $(\pi, \sigma) \in \Gamma \to V_{(\pi,\sigma)} \in \mathscr{U}$ having the asserted properties of the lemma. Define the function $U:X \to \mathscr{U}$ by $U_x = V_{(\pi_x, \sigma_x)}$. U is a Borel map since it is the composition of the Borel map V with the Borel map $x \mapsto (\pi_x, \sigma_x)$, and the relation $\sigma_x = U_x \pi_x U_x^*$ is immediate from the properties of V. □

Note that if A is a separable GCR algebra and if we take $X = \hat{A}_n$ in the above corollary, we obtain a uniqueness statement for the cross sections whose existence was established in 4.1.3.

4.2. Decomposable Operator Algebras

Let (X, μ) be a finite measure space and let $\mathscr{K}$ be a separable Hilbert space, all of which will be fixed throughout this section. In order that certain Hilbert spaces associated with X, μ, $\mathscr{K}$ be separable, we also require that X be countably generated as a Borel space. A vector-valued function $\xi:X \to \mathscr{K}$ is called a Borel function if $(\xi(x), u)$ defines a Borel function on X, for every vector $u \in \mathscr{K}$. The set of all such Borel functions ξ satisfying the condition $\|\xi\|^2 = \int \|\xi(x)\|^2 \, d\mu(x) < \infty$ forms a separable Hilbert space, after we identify functions that agree a.e. (μ), which we denote by $L^2(X, \mu; \mathscr{K})$. Note, for example, that $x \mapsto \|\xi(x)\|$ is a Borel function since it can be expressed as a countable supremum of Borel functions $\sup_n |(\xi(x), v_n)|$, $v_1, v_2, \ldots$ being any countable dense set of unit vectors in $\mathscr{K}$. Let $\mathscr{B}(X)$ denote the set of all complex-valued Borel functions $f: X \to \mathbb{C}$ which are bounded ($\sup_x |f(x)| < \infty$). It is easy to see that each function $f \in \mathscr{B}(X)$ gives rise to a bounded multiplication operator L_f on $L^2(X, \mu; \mathscr{K})$ (namely $(L_f\xi)(x) = f(x)\xi(x)$, $x \in X$, $\xi \in L^2(X, \mu; \mathscr{K})$); such operators are called *diagonal*, and the set $\mathscr{Z}$ of all diagonal operators on $L^2(X, \mu; \mathscr{K})$ forms an abelian $*$-algebra which contains the identity.

In this section we are going to study the family of all separable C^*-algebras $\mathscr{A}$ acting on $L^2(X, \mu; \mathscr{K})$ whose strong closures lie between $\mathscr{Z}$ and $\mathscr{Z}'$. Such a C^*-algebra is called *decomposable*. Note, incidentally, that by 1.2.3, every von Neumann algebra $\mathscr{R}$ such that $\mathscr{Z} \subseteq \mathscr{R} \subseteq \mathscr{Z}'$ is the strong closure of such a C^*-algebra. Considerable information will be obtained about the structure of these algebras, for use later in 4.3. The main results are summarized after Corollary 4 at the end of this section.

An operator valued function $F: X \to \mathscr{L}(\mathscr{K})$ is called a *Borel function* if F becomes measurable when $\mathscr{L}(\mathscr{K})$ is endowed with the Borel structure generated by the weak operator topology; equivalently, $x \mapsto (F(x)u, v)$ is a scalar valued Borel function for every pair of vectors $u, v \in \mathscr{K}$. Let $\mathscr{B}(X, \mathscr{K})$ denote the set of all such functions F which are bounded: $\sup_x \|F(x)\| < \infty$. For each $F \in \mathscr{B}(X, \mathscr{K})$ the real-valued function $\|F(x)\|$ can be written as $\sup_{m,n} |(F(x)v_m, v_n)|$, where $v_1, v_2, \ldots$ is a countable dense set in the unit ball of $\mathscr{K}$, and therefore $\|F(x)\|$ is a Borel function. Similar arguments, along with some juggling with inner products, show that $\mathscr{B}(X, \mathscr{K})$ is closed under (pointwise) sums, products, and adjoints, and that for each $F \in \mathscr{B}(X, \mathscr{K})$ and $\xi \in L^2(X, \mu; \mathscr{K})$, $F(x)\xi(x)$ defines another element of $L^2(X, \mu; \mathscr{K})$.

In particular, the pointwise operations and the norm $\|F\| = \sup_x \|F(x)\|$ make $\mathscr{B}(X, \mathscr{K})$ into a normed $*$-algebra, and in fact $\mathscr{B}(X, \mathscr{K})$ is a C^*-algebra (i.e., is complete and satisfies $\|F^*F\| = \|F\|^2$) with identity. Now let us define, for each $F \in \mathscr{B}(X, \mathscr{K})$, a multiplication operator L_F on $L^2(X, \mu; \mathscr{K})$ as follows: $(L_F\xi)(x) = F(x)\xi(x)$, for $\xi \in L^2(X, \mu; \mathscr{K})$. The above remarks show that $L_F\xi$ belongs to L^2, and the following properties are evident: $\|L_F\| \leqslant \text{ess sup } \|F(x)\| \leqslant \|F\|$, $L_{FG} = L_F L_G$, $L_{aF+bG} = aL_F + bL_G$ for scalars a, b, and $L_{F^*} = (L_F)^*$. Thus $F \mapsto L_F$ is a representation of $\mathscr{B}(X, \mathscr{K})$ on $L^2(X, \mu; \mathscr{K})$. Moreover, it is clear that each operator L_F commutes with the algebra $\mathscr{Z}$ of diagonal operators. The following result shows that in fact these multiplication operators exhaust $\mathscr{Z}'$.

Theorem 4.2.1. *For every operator $T \in \mathscr{Z}'$ there is an $F \in \mathscr{B}(X, \mathscr{K})$ such that $T = L_F$ and $\|F(x)\| \leqslant \|T\|$ for all $x \in X$.*

PROOF. Fix $T \in \mathscr{Z}'$. Now each vector $u \in \mathscr{K}$ can be regarded as a constant function in $L^2(X, \mu; \mathscr{K})$, and so gives rise to a (square integrable) $\mathscr{K}$-valued function $x \mapsto (Tu)(x)$ (of course, Tu is only an equivalence class of functions; what we mean is, choose an element in that class). We claim: for each pair

$u, v \in \mathscr{H}$, $|((Tu)(x), v)| \leqslant \|T\| \cdot \|u\| \cdot \|v\|$ a.e. (μ). Since $((Tu)(x), v)$ is a scalar-valued function in $L^2(\mu)$ this will follow if we prove that

$$\left|\int((Tu)(x), v)f(x)\bar{g}(x)\, d\mu(x)\right| \leqslant \|T\| \cdot \|u\| \cdot \|v\|$$

for all bounded functions f, g satisfying $\|f\|_2, \|g\|_2 \leqslant 1$. For such functions f, g put $\xi(x) = f(x)u$ and $\eta(x) = g(x)v$. Then $\xi = L_f u$ and since L_f belongs to $\mathscr{L}$ and T commutes with $\mathscr{L}$, we have

$$\begin{aligned}\int((Tu)(x), v)f(x)\bar{g}(x)\, d\mu(x) &= \int(f(x)(Tu)(x), g(x)v)\, d\mu(x)\\ &= (L_f Tu, L_g v) = (TL_f u, \eta)\\ &= (T\xi, \eta).\end{aligned}$$

Since $|(T\xi, \eta)| \leqslant \|T\| \cdot \|\xi\| \cdot \|\eta\|$, and since $\|\xi\| \leqslant \|u\|$ and $\|\eta\| \leqslant \|v\|$ (because $\|f\|_2, \|g\|_2 \leqslant 1$), the desired inequality follows.

Now let $u_1, u_2, \ldots$ be a countable dense set in $\mathscr{H}$ and let $\mathscr{H}_0$ be the countable set of all finite sums $\sum_i r_i u_i$, where the r_i's range over the field $\mathbb{Q} + i\mathbb{Q}$ of all complex rational numbers. $\mathscr{H}_0$ is a vector space over $\mathbb{Q} + i\mathbb{Q}$ which is dense in $\mathscr{H}$. Since $\mathscr{H}_0$ is countable, we can find a Borel set $N \subseteq X$ such that $\mu(N) = 0$ and $|((Tu)(x), v)| \leqslant \|T\| \cdot \|u\| \cdot \|v\|$ for all $u, v \in \mathscr{H}_0$ and all $x \in X \backslash N$. Moreover, since T is linear, we must have $T(au + bv)(x) = a(Tu)(x) + b(Tv)(x)$ a.e. (μ), for each $u, v \in \mathscr{H}_0$ and each $a, b \in \mathbb{Q} + i\mathbb{Q}$; because there are only a countable number of such relations, we may assume by throwing out a larger null set $M \supseteq N$ that $T(au + bv)(x) = a(Tu)(x) + b(Tv)(x)$, identically on $X \backslash M$ and simultaneously for all $u, v \in \mathscr{H}_0$, $a, b \in \mathbb{Q} + i\mathbb{Q}$. So for each $x \in X \backslash M$ we have a function $[u, v]_x = ((Tu)(x), v)$ defined on $\mathscr{H}_0 \times \mathscr{H}_0$ having all the properties of a bounded bilinear form which make sense on $\mathscr{H}_0 \times \mathscr{H}_0$. Thus $[u, v]_x$ may be uniquely extended by continuity to a *bona fide* bounded bilinear form on $\mathscr{H} \times \mathscr{H}$, so that a familiar lemma of Riesz provides us with an operator $F(x) \in \mathscr{L}(\mathscr{H})$ such that $(F(x)u, v) = ((Tu)(x), v)$, $u, v \in \mathscr{H}_0$. Now extend the function F to all of X by putting $F(x) = 0$ on M. Clearly $\|F(x)\| \leqslant \|T|$ for all $x \in X$, and the fact that $L_F = T$ is a routine verification which will be left for the reader. □

Corollary 1. *The map $F \to L_F$ is a $*$-homomorphism of $\mathscr{B}(X, \mathscr{H})$ onto $\mathscr{L}'$.*

Note that $L_F = 0$ if and only if $F(x) = 0$ a.e. (μ); thus these functions form a closed ideal $\mathscr{N}$ in $\mathscr{B}(X, \mathscr{H})$ and the map $F \to L_F$ induces an isometric $*$-isomorphism of $\mathscr{B}(X, \mathscr{H})/\mathscr{N}$ onto $\mathscr{L}'$. Moreover, the quotient is identifiable with an obvious L^∞ space of operator-valued functions, thereby giving a concrete representation of elements of $\mathscr{L}'$.

In particular, every C^*-algebra contained in $\mathscr{L}'$ has a replica in $\mathscr{B}(X, \mathscr{H})/\mathscr{N}$, but it will turn out later that this representation is not convenient for our purposes. What we actually need is a generalization of 4.2.1 which provides a cross section for the map $F \mapsto L_F$, which preserves all the algebraic relations. Corollary 2 provides the required "lifting" theorem.

Corollary 2. *Let $\mathscr{A}$ be a separable C^*-subalgebra of $\mathscr{L}'$. Then there is an isometric $*$-homomorphism $\pi:\mathscr{A} \to \mathscr{B}(X, \mathscr{K})$ such that $L_{\pi(T)} = T$ for all $T \in \mathscr{A}$.*

PROOF. Let $\mathscr{S}$ be a countable subset of $\mathscr{A}$ which generates $\mathscr{A}$ as a C^*-algebra. We may assume $\mathscr{S}^* = \mathscr{S}$, and that $\mathscr{S}$ contains the identity if $\mathscr{A}$ does. Let $\mathscr{A}_0$ be the countable family of all finite sume of operators of the form $rA_1A_2 \cdots A_n$, where $A_j \in \mathscr{S}$, $r \in \mathbb{Q} + i\mathbb{Q}$. $\mathscr{A}_0$ is clearly dense in $\mathscr{A}$ and is a $*$-algebra over the field $\mathbb{Q} + i\mathbb{Q}$ of rational complex numbers. For each element $T \in \mathscr{A}_0$, choose a function $F_T \in \mathscr{B}(X, \mathscr{K})$ as in 4.2.1. Now for $a, b \in \mathbb{Q} + i\mathbb{Q}$ and $S, T \in \mathscr{A}_0$, we must have $F_{aS+bT}(x) = aF_S(x) + bF_T(x)$ a.e. (μ) and $(F_{T^*})(x) = (F_T(x))^*$ a.e. (μ) [because $G \mapsto L_G$ is a$*$-homomorphism and $L_G = 0$ iff $G(x) = 0$ a.e. (μ)]. Since there are only a countable number of such relations there is a Borel set N such that $\mu(N) = 0$, and *all* of the above relations are satisfied on $X \backslash N$. Define $\pi:\mathscr{A}_0 \mapsto \mathscr{B}(X, \mathscr{K})$ by $\pi(T)(x) = F_T(x)$ if $x \in X\backslash N$, and $\pi(T)(x) = 0$ if $x \in N$. Then π is a $*$-homomorphism of $\mathscr{A}_0$ into $\mathscr{B}(X, \mathscr{K})$ satisfying $L_{\pi(T)} = T$ and $\|\pi(T)(x)\| \leqslant \|T\|$, $T \in \mathscr{A}_0$, $x \in X$. Note that $\|T\| \leqslant \text{ess sup}\|\pi(T)(x)\| \leqslant \sup\|\pi(T)(x)\|$ follows from the first equation, so that π is in fact isometric. Thus the unique extension of π to $\mathscr{A}$ has all the right properties. □

Let $\mathscr{A}$ be a separable C^*-subalgebra of $\mathscr{L}'$, and choose a homomorphism $\pi:\mathscr{A} \to \mathscr{B}(X, \mathscr{K})$ as in Corollary 2. Then for each $x \in X$, we obtain a representation π_x of $\mathscr{A}$ on $\mathscr{K}$ by evaluating π at $x:\pi_x(T) = \pi(T)(x)$, $T \in \mathscr{A}$. We shall require some information about this family of representations of $\mathscr{A}$.

Proposition 4.2.2. *Let $F, F_1, F_2, \ldots$ be a sequence in $\mathscr{B}(X, \mathscr{K})$, such that the sequence L_{F_n} of operators converges strongly to L_F. Then there is a subsequence $n' \subseteq n$ such that $F_{n'}(x)$ converges strongly to $F(x)$ for μ-almost every $x \in X$.*

PROOF. The idea is simple, and we merely sketch the details. Note first that the norms $\|L_{F_n}\|$ are bounded (Banach–Steinhaus theorem). Letting $v_1, v_2, \ldots$ be a dense sequence in $\mathscr{K}$, we see that

$$\lim_n \int \|F_n(x)v_j - F(x)v_j\|^2 \, d\mu(x) = 0 \quad \text{for each } j = 1, 2, \ldots.$$

Thus for each j there is a subsequence $n_{1j} < n_{2j} < \cdots$ so that

$$\lim_k \|F_{n_{kj}}(x)v_j - F(x)v_j\| = 0$$

for all x in the complement of a μ-null set $N_j \subseteq X$. By a simple induction we can even arrange that the $(k+1)$st sequence is a subsequence of the kth. The required subsequence is obtained by diagonalization (its kth term is the kth term of the kth subsequence), and the exceptional null set can be taken as $N_1 \cup N_2 \cup \cdots$. By discarding another null set we may assume $\|F_n(x)\| \leqslant \|L_{F_n}\|$ for all x and all n, and now the boundedness condition and the density of $\{v_j\}$ in $\mathscr{K}$ insures strong convergence almost everywhere of the subsequence. □

Now let $\mathscr{A} \subseteq \mathscr{Z}'$ be a separable C^*-algebra with trivial nullspace, and let $\pi: \mathscr{A} \to \mathscr{B}(X, \mathscr{K})$ be as in Corollary 2 above.

Corollary 1. *If $F \in \mathscr{B}(X, \mathscr{K})$ is such that $L_F \in \mathscr{A}''$, then $F(x) \in \pi_x(\mathscr{A})''$ for almost every $x \in X$.*

PROOF. It suffices to prove the assertion separately for the real and imaginary parts of F, so we can assume F (and therefore L_F) is self-adjoint.

By the double commutant theorem and the corollary of Kaplansky's density theorem (1.2.2), we can find a sequence $T_n = T_n^* \in \mathscr{A}$ such that $T_n \to L_F$ strongly. By Lemma 1 (and by passing to a subsequence if necessary) we may assume $\pi(T_n)(x) \to F(x)$ strongly, for almost every $x \in X$. Thus $F(x)$ belongs to the strong closure of $\pi_x(\mathscr{A})$ a.e. (μ), as required. □

Recall that two representations π, σ of a C^*-algebra are called disjoint if no nonzero subrepresentation of π is equivalent to any subrepresentation of σ.

Corollary 2. *If $\mathscr{A}$ is strongly dense in $\mathscr{Z}'$, then π_x is irreducible for almost every $x \in X$.*

PROOF. Choose any irreducible operator $T \in \mathscr{L}(\mathscr{K})$, and let F be the constant function $F(x) = T, x \in X$. Then $L_F \in \mathscr{Z}'$, so by hypothesis $L_F \in \mathscr{A}''$. By Corollary 1 we conclude that $T \in \pi_x(\mathscr{A})''$ a.e. (μ), from which the conclusion is evident. □

Corollary 3. *Assume $\mathscr{A}$ has trivial nullspace. If $\mathscr{Z} \subseteq \mathscr{A}'' \subseteq \mathscr{Z}'$, then there is a Borel set $N \subseteq X$ of measure zero such that π_x and π_y are disjoint for all pairs $x \neq y$ in $X \backslash N$.*

PROOF. Since X is countably separated (in fact, countably generated), we can find a sequence $f_1, f_2, \ldots$ of bounded real valued Borel functions on X which separates points. Each multiplication operator L_{f_n} belongs to $\mathscr{Z}$, and since $\mathscr{Z}$ is contained in the strong closure of $\mathscr{A}$ by the double commutant theorem, we may find, for each n, a sequence $A_{n1}, A_{n2}, \ldots$ in $\mathscr{A}$ such that $\lim_k A_{nk} = L_{f_n}$ strongly, for every $n = 1, 2, \ldots$ (imitate the proof of Corollary 1). Since $A_{nk} = L_{\pi(A_{nk})}$ we may assume (utilizing 4.2.2 and passing to subsequences if necessary) that for each n, $\lim_k \pi_x(A_{nk}) = f_n(x)I$ strongly for all x in the complement of some μ-null set N_n. Put $N = N_1 \cup N_2 \cup \cdots$.

Now choose $x \neq y$ in $X \backslash N$. To see that π_x and π_y are disjoint, let T be a bounded operator on $\mathscr{K}$ such that $T\pi_x(A) = \pi_y(A)T$, $A \in \mathscr{A}$. This remains true for $A = A_{nk}$ as above, and thus by taking strong limits on k (which exist on $X \backslash N$) we conclude $Tf_n(x) = f_n(y)T = Tf_n(y)$, for all $n = 1, 2, \ldots$. Since $x \neq y$ there is an n for which $f_n(x) \neq f_n(y)$, and this implies $T = 0$. From 2.1.4 (ii) we conclude that $\pi_x \circ \pi_y$. □

Corollary 3 has a plausible converse. Namely, let $\mathscr{A}$ be a separable C^*-subalgebra of $\mathscr{L}'$ and suppose that we have a map $\pi:\mathscr{A} \to \mathscr{B}(X, \mathscr{K})$ as in Corollary 1, which satisfies $\pi_x \mathrel{\not\sim} \pi_y$ for all $x \neq y$ in the complement of a μ-null set. Then it is quite reasonable to conjecture that $\mathscr{A}''$ contains the set $\mathscr{L}$ of diagonal operators. This can be proved, in fact, if μ is an atomic measure. But in general, and surprisingly perhaps, it is false. It turns out that even when X is a *standard* Borel space, there is a finite Borel measure μ on X, a separable C^*-algebra $\mathscr{A} \subseteq \mathscr{L}'$, and a map $\pi:\mathscr{A} \to \mathscr{B}(X, \mathscr{K})$ as in Corollary 1 all of which satisfy

(i) $\pi_x \mathrel{\not\sim} \pi_y$ *for all* $x \neq y$ *in* X, *and*
(ii) $\mathscr{A}''$ *is a factor* (*not of type* I)

(cf. [8], Corollary 4 et seq.). Because of (ii) there is of course no hope of having $\mathscr{L} \subseteq \mathscr{A}''$; for $\mathscr{L} \subseteq \mathscr{A}'' \subseteq \mathscr{L}'$ implies that $\mathscr{L}$ is contained in the center of $\mathscr{A}''$, which by (ii) is only one-dimensional. This behavior illustrates only one kind of pathology that occurs in the representation theory of C^*-algebras which are not GCR. We will describe others in the next section.

The next result gives a useful and concrete description of operators in $\mathscr{A}''$, when $\mathscr{L} \subseteq \mathscr{A}'' \subseteq \mathscr{L}'$. We assume, of course, that $\mathscr{A}$ has trivial null-space. In terms of direct integrals (which we have deliberately avoided until now) it asserts that $\mathscr{A}''$ is the direct integral $\int^{\oplus} \pi_x(\mathscr{A})''\, d\mu(x)$. More generally, it leads directly to the existence of the central decomposition for von Neumann algebras acting on a separable Hilbert space. However, we will not go into that here.

Corollary 4. *Assume* $\mathscr{L} \subseteq \mathscr{A}'' \subseteq \mathscr{L}'$. *Then* $\mathscr{A}''$ *coincides with the set of all multiplications* L_F, $F \in \mathscr{B}(X, \mathscr{K})$, *satisfying* $F(x) \in \pi_x(\mathscr{A})''$ a.e. (μ).

Proof. The inclusion $\subseteq$ is Corollary 1 above. For the other inclusion choose $F \in \mathscr{B}(X, \mathscr{K})$ satisfying $F(x) \in \pi_x(\mathscr{A})''$ a.e. We will show that L_F commutes with $\mathscr{A}'$. Choose $T \in \mathscr{A}'$. Since $\mathscr{L} \subseteq \mathscr{A}''$ we have $\mathscr{A}' = \mathscr{A}''' \subseteq \mathscr{L}'$ and so by 4.2.1 T has the form L_G for some $G \in \mathscr{B}(X, \mathscr{K})$. Let $A_1, A_2, \ldots$ be a generating sequence for $\mathscr{A}$. Then since L_G commutes with each $A_n = L_{\pi(A_n)}$ we have $G(x)\pi_x(A_n) = \pi_x(A_n)G(x)$ a.e. (μ) for every $n = 1, 2, \ldots$. Thus there is a Borel set N of measure zero such that $G(x)\pi_x(A_n) = \pi_x(A_n)G(x)$ for every $x \in X\backslash N$ and every $n = 1, 2, \ldots$. Hence $G(x) \in \pi_x(\mathscr{A})'$ for all $x \in X\backslash N$, and since $F(x) \in \pi_x(\mathscr{A})''$ a.e. (μ) we conclude that $F(x)$ commutes with $G(x)$ a.e. (μ). It follows that L_F commutes with $L_G = T$. □

Let us now summarize the main points. For every separable C^*-algebra $\mathscr{A} \subseteq \mathscr{L}'$, there is an isometric $*$-homomorphism $\pi:\mathscr{A} \to \mathscr{B}(X, \mathscr{K})$ satisfying $A = L_{\pi(A)}$, $A \in \mathscr{A}$. The representations $\pi_x(A) = \pi(A)(x)$ have the properties:

(i) *If* $\mathscr{A}'' = \mathscr{L}'$ *then* π_x *is irreducible for almost every* $x(\mu)$.

(ii) *If* $\mathscr{Z} \subseteq \mathscr{A}'' \subseteq \mathscr{Z}'$ *then the* $\{\pi_x\}$ *are mutually disjoint for all* x *in the complement of a* μ*-null set.*
(iii) *If* $\mathscr{Z} \subseteq \mathscr{A}'' \subseteq \mathscr{Z}'$ *then* $\mathscr{Z}''$ *consists precisely of those multiplications* L_F *satisfying* $F(x) \in \pi_x(\mathscr{A})''$ a.e. (μ).

4.3. Representations of GCR Algebras

In Section 2.2 we classified the separable representations of separable abelian C^*-algebras in terms of measure classes and multiplicity functions. We are now going to imitate that program for general separable GCR algebras. While the details are somewhat more complicated in the GCR case, the basic steps are the same. First, we will choose a finite Borel measure μ on the spectrum of the algebra and construct a multiplicity-free representation π_μ, which will depend only on the measure class of μ. Second, we will show that every separable multiplicity-free representation is equivalent to one of these π_μ's. Third, we will utilize the results of Chapter 2 to classify general type I representations in terms of the multiplicity-free representations.

Let A be a separable GCR algebra, which will be fixed throughout this section. Choose a finite Borel measure μ on the spectrum $\hat{A}$ of A. Since $\hat{A}$ is the direct sum $\hat{A}_\infty \cup \hat{A}_1 \cup \hat{A}_2 \ldots$, it is convenient to deal with each piece $\hat{A}_n$ separately; thus we will assume for the moment that μ is concentrated on some $\hat{A}_n$, intending to reassemble the pieces later. By 4.1.3 there is a Borel mapping $\zeta \in \hat{A}_n \mapsto \pi_\zeta \in \operatorname{irr}(A, \mathscr{H}_n)$ such that π_ζ belongs to the class ζ, for every $\zeta \in \hat{A}_n$. Now form the "direct integral" $\pi_\mu = \int^\oplus \pi_\zeta \, d\mu(\zeta)$. More precisely, consider the Hilbert space $L^2(\hat{A}_n, \mu; \mathscr{H}_n)$ of all $\mathscr{H}_n$-valued Borel functions $\xi: \hat{A}_n \to \mathscr{H}_n$ such that $\|\xi\|^2 = \int \|\xi(\zeta)\|^2 \, d\mu(\zeta) < \infty$. For every $x \in A$, we may define an operator $\pi_\mu(x)$ on $L^2(\hat{A}_n, \mu; \mathscr{H}_n)$ by

$$(\pi_\mu(x)\xi)(\zeta) = \pi_\zeta(x)\xi(\zeta),$$

$\zeta \in \hat{A}_n$, $\xi \in L^2(\hat{A}_n, \mu; \mathscr{H}_n)$. It is not hard to see that everything in sight is Borel measurable, because $\zeta \mapsto \pi_\zeta$ is a Borel map of $\hat{A}_n$ into $\operatorname{irr}(A, \mathscr{H}_n)$. Thus π_μ is a representation of A on $L^2(\hat{A}_n, \mu; \mathscr{H}_n)$. It is clear that π_μ depends on μ, and conceivable that it also depends on the particular cross section $\zeta \mapsto \pi_\zeta$. We first want to show that the unitary equivalence class of π_μ is independent of the particular cross section. Indeed, if $\zeta \mapsto \sigma_\zeta$ is another Borel map of $\hat{A}_n$ into $\operatorname{irr}(A, \mathscr{H}_n)$ such that $\sigma_\zeta \in \zeta$ for all $\zeta \in \hat{A}_n$, then we may form $\sigma_\mu = \int^\oplus \sigma_\zeta \, d\mu(\zeta)$ in the same way we formed π_μ. We claim: σ_μ and π_μ are equivalent. Indeed by the corollary of 4.1.4 there is a Borel map $\zeta \mapsto V_\zeta$ of $\hat{A}_n$ into the unitary group of $\mathscr{H}_n$ such that $\sigma_\zeta(x)V_\zeta = V_\zeta\pi_\zeta(x)$, $x \in A$, $\zeta \in \hat{A}_n$. Define an operator V on $L^2(\hat{A}_n, \mu; \mathscr{H}_n)$ by $(V\xi)(\zeta) = V_\zeta\xi(\zeta)$, $\xi \in L^2$, $\zeta \in \hat{A}_n$. Since V_ζ is unitary for all ζ it follows that V is unitary, and the relation $\sigma_\zeta V_\zeta = V_\zeta \pi_\zeta$ becomes $\sigma_\mu V = V\pi_\mu$, proving the claim.

Therefore π_μ is well-defined by μ up to equivalence. Moreover, it will be convenient to write ζ in place of π_ζ, where it is understood always that the irreducible representations $x \in A \mapsto \zeta(x)$, $\zeta \in \hat{A}_n$, in fact come from a particular Borel cross section for the quotient map. Thus the direct integral notation for π_μ becomes $\pi_\mu = \int^\oplus \zeta \, d\mu(\zeta)$.

The following theorem is the key step leading to the fact that π_μ is multiplicity-free. While the corresponding result for abelian C^*-algebras (2.2.1) is straightforward, the argument here has to be made with some care.

Theorem 4.3.1. *$\pi_\mu(A)''$ contains the algebra $\mathscr{X}$ of all diagonal operators.*

PROOF. It suffices to show that $\pi_\mu(A)''$ contains all diagonal operators of the form L_f where f is the characteristic function of a Borel set in $\hat{A}_n$. Let $\mathscr{B}_0$ be the class of all Borel sets in $\hat{A}_n$ having this property. Clearly $\mathscr{B}_0$ is closed under complementation, and a simple application of the bounded convergence theorem shows that $\mathscr{B}_0$ is closed under countable unions. Thus $\mathscr{B}_0$ is a sub σ-field of the Borel field $\mathscr{B}$ on $\hat{A}_n$. Since $\hat{A}_n$ is a standard Borel space (corollary to 4.1.3.), the unique structure theorem (3.3.5) shows that $\mathscr{B}_0 = \mathscr{B}$ provided we can find a countable subfamily of $\mathscr{B}_0$ which separates points.

This countable subfamily is constructed as follows. Let $r_1, r_2, \ldots$ be an enumeration of the nonnegative rational real numbers, let $a_1, a_2, \ldots$ be a countable dense set in A, and define subsets $S_{k\ell} \subseteq \text{irr}(A, \mathscr{H}_n)$ by

$$S_{k\ell} = \{\pi \in \text{irr}(A, \mathscr{H}_n) : \|\pi(a_k)\| \leqslant r_\ell\} \qquad k, \ell \geqslant 1.$$

Letting q be the natural map of $\text{irr}(A, \mathscr{H}_n)$ onto $\hat{A}_n$, we obtain a countable family of sets $q(S_{k\ell})$ in $\hat{A}_n$. Now each $S_{k\ell}$ is clearly closed in $\text{irr}(A, \mathscr{H}_n)$, and is saturated with respect to the equivalence relation $\sim$. Thus each image $q(S_{k\ell})$ is a Borel set in $\hat{A}_n$. We claim that $\{q(S_{k\ell})\}$ separates points in $\hat{A}_n$; equivalently, if $\pi, \sigma \in \text{irr}(A, \mathscr{H}_n)$ are such that $\pi \in S_{k\ell}$ iff $\sigma \in S_{k\ell}$ for all $k, \ell = 1, 2, \ldots$, then $\pi \sim \sigma$. To see this, note that the hypothesis means $\|\pi(a_k)\| \leqslant r$ iff $\|\sigma(a_k)\| \leqslant r$ for every rational $r \geqslant 0$ and every $k = 1, 2, \ldots$, and hence $\|\pi(a_k)\| = \|\sigma(a_k)\|$ for all k. Since $\{a_1, a_2, \ldots\}$ is dense in A we conclude $\|\pi(x)\| = \|\sigma(x)\|$ for all $x \in A$ and in particular π and σ have the same kernel. By 1.5.4, π and σ are equivalent, as asserted.

It remains to prove that each set $q(S_{k\ell})$ belongs to $\mathscr{B}_0$. For that, fix k and ℓ, and let J be the ideal $\bigcap \{\ker \sigma : \sigma \in S_{k\ell}\}$. First, we claim $S_{k\ell} = \{\pi \in \text{irr}(A, \mathscr{H}) : \pi(J) = 0\}$. Let τ be the (possibly inseparable) representation of A defined by $\tau = \bigoplus \sigma, \sigma \in S_{k\ell}$. Since J is the kernel of τ and $\pi(J) = 0$, it follows that the map $\tau(x) \mapsto \pi(x)$, $x \in A$, is a well-defined representation of the C^*-algebra $\tau(A)$, and so by 1.3.2 we have $\|\pi(x)\| \leqslant \|\tau(x)\| = \sup_{\sigma \in Sk\ell} \|\sigma(x)\|$ for all $x \in A$. In particular this inequality shows that $\|\pi(a_k)\| \leqslant r_\ell$, hence $\pi \in S_{k\ell}$, proving the claim.

Now let $e_1, e_2, \ldots$ be an approximate identity for the ideal J (Section 1.3). It is clear that if $\sigma \in S_{k\ell}$ then $\sigma(e_j) = 0$ for all j (because $\sigma(J) = 0$). Conversely, suppose $\sigma \in \text{irr}(A, \mathscr{H}_n) \backslash S_{k\ell}$. Then we claim: $\sigma(e_j) \to I$ in the strong operator topology of $\mathscr{L}(\mathscr{H}_n)$, as $j \to \infty$. Indeed, since $\|\sigma(e_j)\| \leqslant 1$ it suffices to show that $\|\sigma(e_j)\eta - \eta\| \to 0$ for all η in a dense set of $\mathscr{H}_n$. Now by the preceding paragraph $\sigma(J) \neq 0$, and since J is an ideal $\sigma(J)$ must therefore be an irreducible C^*-algebra in $\mathscr{L}(\mathscr{H}_n)$ (1.3.4). Thus $\sigma(J)\xi$ is dense for every

$\xi \neq 0$. But if η has the form $\sigma(z)\xi$ with $z \in J$ then $\sigma(e_j)\eta - \eta = \sigma(e_jz - z)\eta \to 0$ since $\|e_jz - z\| \to 0$, proving the claim.

Finally, let f be the characteristic function of $\hat{A}_n \backslash q(S_{k\ell})$. We want to show that L_f belongs to $\pi_\mu(A)''$. Now for each j, we obtain a bounded $\mathscr{L}(\mathscr{H}_n)$-valued Borel function E_j on $\hat{A}_n$ defined by $E_j(\zeta) = \zeta(e_j)$, $\zeta \in \hat{A}_n$. It is clear that the multiplication operator L_{E_j} belongs to $\pi_\mu(A)$ (L_{E_j} is just $\pi_\mu(e_j)$). Moreover, the preceding paragraph shows that $E_j(\zeta) \to f(\zeta)I$ strongly, for each $\zeta \in \hat{A}_n$. An application of the bounded convergence theorem shows that $L_{E_j} \to L_f$ strongly in $L^2(\hat{A}_n, \mu; \mathscr{H}_n)$, and in particular we conclude that $L_f \in \pi_\mu(A)''$. □

Corollary 1. $\pi_\mu(A)'' = \mathscr{Z}'$.

PROOF. It is obvious that $\pi_\mu(A)$ commutes with $\mathscr{Z}$, so by the above theorem we have $\mathscr{Z} \subseteq \pi_\mu(A)'' \subseteq \mathscr{Z}'$. Moreover, since the mapping which associates with each operator $\pi_\mu(x)$ in $\pi_\mu(A)$ the Borel function $F(\zeta) = \zeta(x)$ satisfies the hypothesis of Corollary 4 of 4.2.2., we see that $\pi_\mu(A)''$ coincides with the family of all multiplications L_F, where $F \in \mathscr{B}(\hat{A}_n, \mu; \mathscr{H}_n)$ satisfies $F(\zeta) \in \zeta(A)''$ a.e. (μ). Since $\zeta(A)'' = \mathscr{L}(\mathscr{H}_n)$ for every $\zeta \in \hat{A}_n$ (by irreducibility) we conclude that $\pi_\mu(A)''$ contains all multiplication operators. The conclusion now follows from 4.2.1.

Corollary 2. *π_μ is multiplicity-free.*

PROOF. By Corollary 1, $\pi_\mu(A)'$ is the abelian von Neumann algebra $\mathscr{Z}''$. □

We remark that Theorem 4.3.1 and its corollaries are false if A is not GCR. Indeed, for every separable non-GCR algebra A one can find a *standard* Borel space X (which can be taken as a Borel subset of $\hat{A}_\infty$), a measure μ on X, and a Borel map $x \mapsto \pi_x$ of X into irr$(A, \mathscr{H}_n)$, such that π_x and π_y are disjoint for $x \neq y$, and which has the property that the range of the representation $\int^\oplus \pi_x \, d\mu(x)$ generates a *factor not of type I* (cf. [8] and the discussion following Corollary 3 of 4.2.2.). In particular $\int^\oplus \pi_x \, d\mu(x)$ is *not* multiplicity-free. We have already remarked in 4.1 that the spectrum of a non-GCR algebra is "bad" as a Borel space. The above phenomenon illustrates quite a different source of pathology associated with those C^*-algebras.

Note also that Corollary 1 implies that the center of the von Neumann algebra $\pi(A)''$ coincides with the center of $\mathscr{Z}'$, which is in turn the von Neumann algebra $\mathscr{Z}$ of all diagonal operators. Since μ is concentrated on $\hat{A}_n$, $\mathscr{Z}'$ is identified with the set of all multiplication operators L_F, $F \in \mathscr{B}(\hat{A}_n, \mu; \mathscr{H}_n)$; and since dim $\mathscr{H}_n = n$, we see that the identity representation of $\mathscr{Z}$ has multiplicity n. Conclusion: *if μ is concentrated on $\hat{A}_n$, then the center of $\pi_\mu(A)''$ has multiplicity n.* This tells us how to find n, given π_μ and the knowledge that μ is concentrated on some particular piece $\hat{A}_n$ of $\hat{A}$.

Now, still assuming that μ is concentrated on $\hat{A}_n$, we want to investigate how the multiplicity-free representation π_μ depends on μ. The answer is the same as in the abelian case.

Theorem 4.3.2. *Let μ and ν be two finite Borel measures concentrated on $\hat{A}_n$. Then $\pi_\mu \sim \pi_\nu$ iff μ and ν are equivalent measures, and $\pi_\mu \circ\!\!\!\!\!\!\smile \pi_\nu$ iff μ and ν are mutually singular.*

PROOF. We have arranged the proof of 2.2.2, so that it translates verbatim to a proof of this theorem, provided one uses 4.3.1. in place of 2.2.1. For example, to see that $\mu \sim \nu$ implies $\pi_\mu \sim \pi_\nu$, let $h = d\mu/d\nu$ as in 2.2.2., and define $U: L^2(\hat{A}_n, \mu; \mathscr{H}_n) \to L^2(\hat{A}_n, \nu; \mathscr{H}_n)$ by the same formula as in 2.2.2

$$(U\xi)(\zeta) = h(\zeta)^{1/2}\xi(\zeta) \qquad \zeta \in \hat{A}_n,\ \xi \in L^2(\hat{A}_n, \mu; \mathscr{H}_n)$$

(this makes perfect sense even though ξ is vector-valued). Then we see that

$$\int \|U\xi(\zeta)\|^2\, d\nu(\zeta) = \int \|\xi(\zeta)\|^2 h(\zeta)\, d\nu(\zeta) = \int \|\xi(\zeta)\|^2\, d\mu(\zeta),$$

showing that U is an isometry, and the arguments for $UU^* = I$ and $U\pi_\mu = \pi_\nu U$ are completed in the same way. □

Now let us put the pieces together. Namely, choose a finite Borel measure μ on $\hat{A}$, where we do not assume that μ is concentrated on some $\hat{A}_n$. If μ_n is the restriction of μ to $\hat{A}_n$, then we can form π_{μ_n} as above, obtaining a separable multiplicity-free representation. Define $\pi_\mu = \pi_{\mu_\infty} \oplus \pi_{\mu_1} \oplus \pi_{\mu_2} \oplus \cdots$. One can also think of π_μ as acting on a space of vector-valued functions. Briefly, let $L^2(\hat{A}, \mu; \mathscr{H}_\infty \cup \mathscr{H}_1 \cup \mathscr{H}_2 \cup \cdots)$ denote the space of all Borel functions ξ from $\hat{A}$ into the direct sum $\bigcup_n \mathscr{H}_n$ of the Borel spaces $\mathscr{H}_n$, such that $\xi(\zeta) \in \mathscr{H}_n$ for all $\zeta \in \hat{A}_n$ and for which the norm $\|\xi\| = (\int_A \|\xi(x)\|^2\, d\mu(x))^{1/2}$ is finite. Then for each $x \in A$, $\pi_\mu(x)$ acts as follows: $(\pi_\mu(x)\xi)(\zeta) = \zeta(x)\xi(\zeta)$, for every $\zeta \in \hat{A}_n$ and every $n = \infty, 1, 2, \ldots$ (the reader may wish to examine this realization of π_μ in more detail). In order to suggest this interpretation, we will write $\pi_\mu = \int_{\hat{A}}^{\oplus} \zeta\, d\mu(\zeta)$. In any case, π_μ is clearly a separable representation (because each π_{μ_n} is), and we want to show now that π_μ is multiplicity-free.

Proposition 4.3.3. *Let μ and $\pi_\mu = \pi_{\mu_\infty} \oplus \pi_{\mu_1} \oplus \pi_{\mu_2} \oplus \cdots$ be as above, and let $P_\infty, P_1, P_2, \ldots$ be the projections of the Hilbert space $\mathscr{H}$ of π_μ onto the respective coordinate subspaces. Let $\mathscr{C}$ be the center of $\pi_\mu(A)''$. Then each $P_n \in \mathscr{C}$, π_μ is multiplicity-free, and the abelian von Neumann algebra $\mathscr{C}|_{P_n\mathscr{H}}$ is 0 or has multiplicity n, for every $n = \infty, 1, 2, \ldots$.*

PROOF. We claim first than π_{μ_m} and π_{μ_n} are disjoint if $m \neq n$. To prove this, it suffices to show that for every nonzero subrepresentation σ of π_{μ_n}, $\sigma(A)'$ is an abelian von Neumann algebra having multiplicity n (this clearly implies that no nonzero subrepresentation of π_{μ_m} can be equivalent to a subrepresentation of π_{μ_n}). Choose such a $\sigma \leqslant \pi_{\mu_n}$. By Corollary 1 of 4.3.1, the range projection of σ belongs to $\mathscr{C}$, and is therefore multiplication by the characteristic function of a Borel set $E \subseteq \hat{A}_n$. Letting ν be the measure on $\hat{A}_n$ defined by $\nu(S) = \mu_n(S \cap E)$, we have $\sigma = \pi_\nu$, and the assertion now follows from 4.3.1 and the remark preceding 4.3.2.

Since each subrepresentation π_{μ_n} of π_μ is multiplicity-free, we can now deduce from 2.1.6 (ii) that π_μ is multiplicity-free. Therefore $\pi_\mu(A)' = \mathscr{C}$, and since each P_n clearly commutes with $\pi_\mu(A)$, we see that P_n belongs to $\mathscr{C}$.

To prove the last assertion of the theorem we claim that $\mathscr{C}|_{P_n\mathscr{H}}$ is the center of $\pi_{\mu_n}(A)''$; the remark preceding 4.3.2 will then give the desired conclusion. Clearly $\mathscr{C}|_{P_n\mathscr{H}}$ is contained in the commutant of $\pi_{\mu_n}(A)$ (which is, of course, the center of its weak closure). Conversely, if T is an operator which commutes with $\pi_{\mu_n}(A)$, then TP_n commutes with $\pi_\mu(A)$. Hence TP_n belongs to $\mathscr{C}$, and finally $T \in \mathscr{C}|_{P_n\mathscr{H}}$. □

It is now a simple matter to see that the results of 4.3.2 persist for general measures μ on $\hat{A}$, not necessarily concentrated on a single piece $\hat{A}_n$:

Corollary. *Let μ and ν be two finite Borel measures on $\hat{A}$. Then $\pi_\mu \sim \pi_\nu$ iff $\mu \sim \nu$, and $\pi_\mu \,\delta\, \pi_\nu$ iff $\mu \perp \nu$.*

Proof. Assume first that $\pi_\mu \sim \pi_\nu$, and let U be a unitary operator such that $\pi_\nu = U\pi_\mu U^*$. Let $P_\infty, P_1, P_2, \ldots$ (resp. $Q_\infty, Q_1, Q_2, \ldots$) be the projections associated with the decomposition $\pi_\mu = \pi_{\mu\infty} \oplus \pi_{\mu 1} \oplus \cdots$ (resp. $\pi_\nu = \pi_{\nu_\infty} \oplus \pi_{\nu_1} \oplus \cdots$) as in 4.3.3. Then $T \mapsto UTU^*$ maps the center of $\pi_\mu(A)''$ onto the center of $\pi_\nu(A)''$ in a multiplicity-preserving fashion. By 4.3.3 and the uniqueness assertion of 2.1.8 we conclude that $UP_nU^* = Q_n$, $n = \infty, 1, 2, \ldots$. In particular, the restriction of U to the range of P_n implements an equivalence between π_{μ_n} and π_{ν_n}, $n = \infty, 1, 2, \ldots$. By 4.3.2 if follows that $\mu_n \sim \nu_n$, and hence $\mu = \sum_n \mu_n$ is equivalent to $\nu = \sum_n \nu_n$.

Conversely, if $\mu \sim \nu$ and $\mu_n(E) = \mu(\hat{A}_n \cap E)$ (resp. $\nu_n(E) = \nu(\hat{A}_n \cap E)$), then $\mu_n \sim \nu_n$; by 4.3.2 we have $\pi_{\mu_n} \sim \pi_{\nu_n}$ and hence $\pi_\mu = \pi_{\mu_\infty} \oplus \pi_{\mu_1} \oplus \cdots$ is equivalent to $\pi_\nu = \pi_{\nu_\infty} \oplus \pi_{\nu_1} \oplus \cdots$.

If $\pi_\mu \,\delta\, \pi_\nu$ and we define μ_n and ν_n as in the preceding paragraph, then π_{μ_n} and π_{ν_n} are subrepresentations of π_μ and π_ν respectively, so that $\pi_{\mu_n} \,\delta\, \pi_{\nu_n}$. 4.3.2 implies $\mu_n \perp \nu_n$, and since $\mu_n \perp \nu_m$ for $m \neq n$ (because $\hat{A}_n \cap \hat{A}_m = \varnothing$) we see that $\mu_n \perp \nu = \sum_n \nu_n$, hence $\mu = \sum_n \mu_n \perp \nu$.

Finally, assume $\mu \perp \nu$, and let μ_n, ν_n be as above. Then $\mu_n \perp \nu_n$ for each $n = \infty, 1, 2, \ldots$, so by 4.3.2 we have $\pi_{\mu_n} \,\delta\, \pi_{\nu_n}$. If $m \neq n$ then we also have $\pi_{\mu_m} \,\delta\, \pi_{\nu_n}$; for 4.3.3 applied to the measure $\mu_n + \nu_m$ implies that $I \oplus 0$ belongs to the center of $\pi_{\mu_n} \oplus \pi_{\nu_m}(A)''$, and as in the proof of 2.2.2 this implies $\pi_{\mu_n} \,\delta\, \pi_{\nu_m}$. It follows that each π_{μ_n} is disjoint from $\pi_\nu = \pi_{\nu_\infty} \oplus \pi_{\nu_1} \oplus \cdots$ (a fact easily seen by considering the intertwining space of $\pi_{\nu_\infty} \oplus \pi_{\nu_1} \oplus \cdots$ and π_{μ_n}) and similarly $\pi_\mu = \pi_{\mu_\infty} \oplus \pi_{\mu_1} \oplus \cdots$ is disjoint from π_ν. □

Thus we conclude, as in the commutative case, that the unitary equivalence classes of the multiplicity-free representation $\pi_\mu = \int_{\hat{A}}^{\oplus} \zeta \, d\mu(\zeta)$ correspond 1—1 with the (finite) measure classes on $\hat{A}$.

We will now prove that every multiplicity-free representation "is" one of these π_μ.

Theorem 4.3.4. *For every nondegenerate multiplicity-free representation π of A, there is a finite Borel measure μ on $\hat{A}$ such that π is equivalent to π_μ.*

PROOF. Let π be a multiplicity-free representation of A on a Hilbert space $\mathscr{H}$. Note that since A is separable, 2.2.3 implies that $\mathscr{H}$ is a separable space. Let us first make a reduction. Decompose the center $\mathscr{C}$ of $\pi(A)''$ as in 2.1.8. That is, find orthogonal projections $C_\infty, C_1, C_2, \ldots$ in $\mathscr{C}$ such that $C_\infty + C_1 + C_2 + \cdots = I$ and such that $\mathscr{C}|_{C_n}\mathscr{H}$ has multiplicity n. Let π_n be the restriction of π to $C_n\mathscr{H}$. Then $\pi = \pi_\infty \oplus \pi_1 \oplus \pi_2 \oplus \cdots$, and if we can find measures μ_n concentrated on $\hat{A}_n$ such that π_n is equivalent to π_{μ_n}, then by the discussion preceding 4.3.3 π itself will have the required form.

Therefore we can assume $\mathscr{C}$ has multiplicity n. Now by Proposition 1.2.3 $\mathscr{C}$ contains a separable weakly dense C^*-subalgebra $\mathscr{C}_0$ which we may as well assume contains the identity. Then the spectrum Z of $\mathscr{C}_0$ is a compact metric space (therefore Polish), and the inverse Gelfand map gives an isometric representation ϕ of $C(Z)$ whose range is $\mathscr{C}_0$. Now since $\mathscr{C}_0 = \phi(C(Z))$ is weakly dense in $\mathscr{C}$ and since $\mathscr{C}$ has multiplicity n, it follows that the representation ϕ has multiplicity n. Thus if we let $\mathscr{H}_n$ be the canonical n-dimensional Hilbert space chosen for the construction of $\hat{A}_n$ in Section 4.1, then the concluding discussion in Section 4.1 shows that there is a finite Borel measure v on Z such that ϕ is unitarily equivalent to the "multiplication" representation $f \mapsto L_f$ of $C(Z)$ on the space of vector-valued functions $L^2(Z, v; \mathscr{H}_n)$. Thus we may as well take ϕ to be of this form.

As in Section 4.2, let $\mathscr{B}(Z)$ (resp. $\mathscr{B}(Z, \mathscr{H}_n)$) denote the C^*-algebra of all bounded Borel functions (resp. bounded $\mathscr{L}(\mathscr{H}_n)$-valued Borel functions) on Z, and let $\mathscr{Z} = \{L_f : f \in \mathscr{B}(Z)\}$ be the algebra of all diagonal operators on $L^2(Z, v; \mathscr{H}_n)$. Note that $\mathscr{Z}$ is the weak closure of $\phi(C(Z))$ (by 4.1.1.) and therefore $\mathscr{Z} = \mathscr{C}$ is the commutant of $\pi(A)$.

We want to associate Z with a Borel subset of $\hat{A}_n$ in a suitable way. This is done as follows. By Corollary 2 of 4.2.1. we may choose an isometric $*$-homomorphism ω of the separable C^*-algebra $\pi(A)$ into $\mathscr{B}(Z, \mathscr{H}_n)$ such that, for each $x \in A$, $\pi(x)$ is "multiplication by" the function $\omega \circ \pi(x)$. Thus for each $z \in Z$ we obtain a representation ω_z of $\pi(A)$ on $\mathscr{H}_n$ by $\omega_z(\pi(x)) = (\omega \circ \pi(x))(z)$. Since $\pi(A)$ is strongly dense in $\mathscr{Z}'$ we see by Corollary 2 of 4.2.2. that v-almost every ω_z is irreducible. Thus by discarding a Borel set of v-measure zero, we can assume every ω_z is irreducible (note that after this deletion Z is no longer a Polish space, but it does remain a *standard* Borel space). Also, $\mathscr{Z} \subseteq \mathscr{Z}' = \pi(Z)''$ and Corollary 3 of 4.2.2. shows that by discarding another Borel null set from Z we may assume ω_{z_1} and ω_{z_2} are inequivalent for all $z_1 \neq z_2$ in Z. Now consider the function $z \mapsto \omega_z \circ \pi$. This maps Z into $\text{irr}(A, \mathscr{H}_n)$, different points go into inequivalent representations, and a simple check (which we omit) shows that it is in fact a Borel map. Thus, letting q be the canonical map of $\text{irr}(A, \mathscr{H}_n)$ into $\hat{A}_n$, we see that $z \mapsto q \circ \omega_z \circ \pi$ is a 1—1 Borel map of the standard Borel space Z into the standard Borel space $\hat{A}_n$. By 3.3.2. (regarding $\hat{A}_n$ as a Borel subset of a Polish space, by definition of standard Borel spaces) it follows that the range of this map is a Borel set in $\hat{A}_n$ and that the function itself is a *Borel isomorphism* of Z with a Borel set in $\hat{A}_n$. Note that in this identification, $z \mapsto \omega_z \circ \pi$ becomes a Borel cross section (restricted to Z) for the quotient map $q:\text{irr}(A, \mathscr{H}_n) \to \hat{A}_n$. Finally, define a Borel measure μ_n on $\hat{A}_n$ by $\mu_n(E) = \nu(Z \cap E)$. Therefore, this change of variables has converted each operator $\pi(x)$, $x \in A$, into the operator $\pi_{\mu_n}(x)$ on $L^2(\hat{A}_n, \mu_n; \mathscr{H}_n)$ defined by $(\pi_{\mu_n}(x)\xi)(z) = \omega_z \circ \pi(x)\xi(z)$, for $z \in Z$, 0 for $z \in \hat{A}_n \backslash Z$, where ξ runs over $L^2(\hat{A}_n, \mu_n; \mathscr{H}_n)$. In other words, π is equivalent to $\pi_{\mu_n} = \int_{\hat{A}_n}^{\oplus} \zeta \, d\mu_n(\zeta)$. □

Let us recall now that every nondegenerate representation of a GCR algebra is type I (2.4.1). So if that result is combined with the decomposition theorem (2.1.8) and the preceding analysis, we see that the most general (separable, nondegenerate) representation of A is obtained as follows. Choose a sequence $\mu_\infty, \mu_1, \mu_2, \ldots$ of finite Borel measure on $\hat{A}$ such that $\mu_i \perp \mu_j$ if $i \neq j$. We do *not* assume μ_n is concentrated on $\hat{A}_n$, and in fact there is no connection between these μ_n and the μ_n of the discussion preceding 4.3.3. For each n, let π_n be the corresponding multiplicity-free representation $\int_{\hat{A}}^{\oplus} \zeta \, d\mu_n(\zeta)$, and then define π by

$$\pi = \infty \cdot \pi_\infty \oplus 1 \cdot \pi_1 \oplus 2 \cdot \pi_2 \oplus \cdots .$$

Then π is a typical representation of A. Moreover, if $\nu_\infty, \nu_1, \nu_2, \ldots$ is another such choice of measures on $\hat{A}_n$ giving rise in the same way to a representation σ, then π is equivalent to σ iff $\mu_n \sim \nu_n$ for every $n = \infty, 1, 2, \ldots$, and $\pi \,\delta\, \sigma$ iff each μ_n is singular with respect to each ν_m.

As in the commutative case, there is a more revealing description of the representation $\infty \cdot \pi_\infty \oplus 1 \cdot \pi_2 \oplus 2 \cdot \pi_2 \oplus \cdots$ obtained from the set of measures $\{\mu_\infty, \mu_1, \mu_2, \ldots\}$. As in 2.2, we may renormalize each μ_n so that $\mu = \mu_\infty + \mu_1 + \mu_2 + \cdots$ is a finite measure, and we can find a partition $\{E_\infty, E_1, E_2, \ldots\}$ of $\hat{A}$ into disjoint Borel sets so that $\mu_j(S) = \mu(S \cap E_j)$, $j = \infty, 1, 2, \ldots$. Thus we can define a Borel measurable multiplicity function $m:\hat{A} \to \{\infty, 1, 2, \ldots\}$ exactly as in 2.2: $m(\zeta) = k$ if $\zeta \in E_k$. m is unique to a μ-equivalence, and conversely the pair (μ, m) determines uniquely the set of measures $\{\mu_\infty, \mu_1, \mu_2, \ldots\}$ as in 2.2. Let us rewrite the representation associated with (μ, m) symbolically as

$$\pi = \int_{\hat{A}}^{\oplus} m(\zeta) \cdot \zeta \, d\mu(\zeta).$$

Conclusion. *Every separable, nondegenerate representation of A is equivalent to one these, and moreover $\int_{\hat{A}}^{\oplus} m(\zeta) \cdot \zeta \, d\mu(\zeta)$ is equivalent to $\int_{\hat{A}}^{\oplus} n(\zeta) \cdot \zeta \, d\nu(\zeta)$ if and only if μ and ν are equivalent measures and the multiplicity functions m and n agree almost everywhere. These representations are disjoint iff μ and ν are mutually singular.* Now of course the concrete realization of π suggested by the direct integral notation has not been established, but it is a simple matter to do so. In the functional realization that we *have* described above, one simply has to increase the multiplicity of ζ by a factor of k uniformly over the set $E_k = \{\zeta \in \hat{A} : m(\zeta) = k\}$, by taking a direct sum of k copies of the Hilbert space associated with ζ. The details are an unexciting variation on what we did at the end of Section 2.2, and are left for the interested reader.

In the case of commutative algebras A, the classification of the representations of A in Section 2.2 led directly to a complete set of unitary invariants for separably acting normal operators in terms of measure classes on the spectrum of the operator. Thus it is natural to expect that the results of this section will lead to a similar classification of separably acting GCR operators T in terms of measure classes on some sort of "generalized" spectrum of T (recall that T is a GCR operator if $C^*(T)$ is a GCR algebra). All of this is true, and readers who have come this far with us may wish to try their hand at formulating the appropriate theorem for themselves.

Bibliography

1. Arveson, W. B. Unitary invariants for compact operators. *Bull. Am. Math. Soc.*, *76*: 88–91, 1970.
2. ——— Subalgebras of C^*-algebras. *Acta Math.*, *123*: 141–224, 1969.
3. Blackwell, D. *Proc. Nat. Acad. Sci. U.S.A.*, *58*: 1836–1837, 1967.
4. Bourbaki, N. *Elements of Mathematics, General Topology*, Part 2. Reading, Mass.: Addison-Wesley, 1966.
5. Deckard, D. and Pearcy, C. On unitary equivalence of Hilbert–Schmidt operators. *Proc. Am. Math. Soc.*, *16*: 671–675, 1965.
6. Dixmier, J. *Les C^*-algébres et leurs representations.* Paris: Gauthier-Villars, 1964.
7. ——— *Les algebres d'operateurs dans l'espace Hilbertien.* Paris: Gauthier-Villars, 1957.
8. ——— Utilisation des facteurs hyperfinis dan la theorie des C^*-algebres. *C.R. Acad. Sci., Paris*, *258*: 4184–4187, 1964.
9. ——— Dual et quasi-dual d'une algebre de Banach involutive. *Trans. Am. Math. Soc.*, *104*: 278–283, 1962.
10. ——— Sur les structures Boréliennes du spectre d une C^*-algebre. *Publ. Inst. Hautes Etudes Sc.*, No. 6: 297–303, 1960.
11. ——— Sur les C^*-algèbres. *Bull. Soc. Math. France*, *88*: 95–112, 1960.
12. Fell, J. M. G. C^*-algebras with smooth dual. *Ill. J. Math.*, : 221–230, 1960.
13. Glimm, J. Type I C^*-algebras, *Ann. Math.*, 73: 572–612, 1961.
14. Halmos, P. R. *A Hilbert Space Problem Book.* Princeton: Van Nostrand; New York: Springer, 1967.
15. Kaplansky, I. The structure of certain operator algebras. *Trans. Am. Math. Soc.*, *70*: 219–255, 1951.
16. ———A theorem on rings of operators. *Pac. J. Math.*, *1*: 227–232, 1951.
17. ———Projections in Banach algebras. *Ann. Math.*, *53*: 235–249, 1951.
18. Kelley, J. L. *General Topology.* Princeton Van Nostrand; New York: Springer, 1955.

19. Kuratowski, K. *Topology*, Vol. I. New York: Academic Press, 1966.

20. Mackey, G. Borel structures in groups and their duals. *Trans. Am. Math. Soc.*, *85*: 134–165, 1957.

21. von Neumann, J. On rings of operators: Reduction theory. *Ann. Math.*, *50*: 401–485, 1949.

22. Pearcy, C. A complete set of unitary invariants for operators generating finite W^*-algebras of type I. *Pac. J. Math.*, *12*: 1405–1416, 1962.

23. Rickart, C. E. *Banach Algebras*. Princeton: Van Nostrand, 1960.

24. Sakai, S. On type I C^*-algebras. *Proc. Am. Math. Soc.*, *18*: 861–863, 1967.

25. ———On a characterization of type I C^*-algebras. *Bull. Am. Math. Soc.*, *72*: 508–512, 1966.

26. Specht, W. Zur Theorie der MatrizenII. *Jahresbericht der Deutschen Mathematiker Vereinigung*, *50*: 19–23, 1940.

27. Riesz, F. and Sz-Nagy, B. *Functional Analysis*. New York: Frederick Ungar, 1955.

Index

Graduate Texts in Mathematics

continued from page ii

61 Whitehead. Elements of Homotopy Theory.
62 Kargapolov/Merlzjakov. Fundamentals of the Theory of Groups.
63 Bollobas. Graph Theory.
64 Edwards. Fourier Series. Vol. I 2nd ed.
65 Wells. Differential Analysis on Complex Manifolds. 2nd ed.
66 Waterhouse. Introduction to Affine Group Schemes.
67 Serre. Local Fields.
68 Weidmann. Linear Operators in Hilbert Spaces.
69 Lang. Cyclotomic Fields II.
70 Massey. Singular Homology Theory.
71 Farkas/Kra. Riemann Surfaces. 2nd ed.
72 Stillwell. Classical Topology and Combinatorial Group Theory. 2nd ed.
73 Hungerford. Algebra.
74 Davenport. Multiplicative Number Theory. 2nd ed.
75 Hochschild. Basic Theory of Algebraic Groups and Lie Algebras.
76 Iitaka. Algebraic Geometry.
77 Hecke. Lectures on the Theory of Algebraic Numbers.
78 Burris/Sankappanavar. A Course in Universal Algebra.
79 Walters. An Introduction to Ergodic Theory.
80 Robinson. A Course in the Theory of Groups. 2nd ed.
81 Forster. Lectures on Riemann Surfaces.
82 Bott/Tu. Differential Forms in Algebraic Topology.
83 Washington. Introduction to Cyclotomic Fields. 2nd ed.
84 Ireland/Rosen. A Classical Introduction to Modern Number Theory. 2nd ed.
85 Edwards. Fourier Series. Vol. II. 2nd ed.
86 van Lint. Introduction to Coding Theory. 2nd ed.
87 Brown. Cohomology of Groups.
88 Pierce. Associative Algebras.
89 Lang. Introduction to Algebraic and Abelian Functions. 2nd ed.
90 Brøndsted. An Introduction to Convex Polytopes.
91 Beardon. On the Geometry of Discrete Groups.
92 Diestel. Sequences and Series in Banach Spaces.
93 Dubrovin/Fomenko/Novikov. Modern Geometry—Methods and Applications. Part I. 2nd ed.
94 Warner. Foundations of Differentiable Manifolds and Lie Groups.
95 Shiryaev. Probability. 2nd ed.
96 Conway. A Course in Functional Analysis. 2nd ed.
97 Koblitz. Introduction to Elliptic Curves and Modular Forms. 2nd ed.
98 Bröcker/tom Dieck. Representations of Compact Lie Groups.
99 Grove/Benson. Finite Reflection Groups. 2nd ed.
100 Berg/Christensen/Ressel. Harmonic Analysis on Semigroups: Theory of Positive Definite and Related Functions.
101 Edwards. Galois Theory.
102 Varadarajan. Lie Groups, Lie Algebras and Their Representations.
103 Lang. Complex Analysis. 3rd ed.
104 Dubrovin/Fomenko/Novikov. Modern Geometry—Methods and Applications. Part II.
105 Lang. $SL_2(\mathbf{R})$.
106 Silverman. The Arithmetic of Elliptic Curves.
107 Olver. Applications of Lie Groups to Differential Equations. 2nd ed.
108 Range. Holomorphic Functions and Integral Representations in Several Complex Variables.
109 Lehto. Univalent Functions and Teichmüller Spaces.
110 Lang. Algebraic Number Theory.
111 Husemöller. Elliptic Curves.
112 Lang. Elliptic Functions.
113 Karatzas/Shreve. Brownian Motion and Stochastic Calculus. 2nd ed.
114 Koblitz. A Course in Number Theory and Cryptography. 2nd ed.
115 Berger/Gostiaux. Differential Geometry: Manifolds, Curves, and Surfaces.
116 Kelley/Srinivasan. Measure and Integral. Vol. I.
117 Serre. Algebraic Groups and Class Fields.
118 Pedersen. Analysis Now.

119 ROTMAN. An Introduction to Algebraic Topology.
120 ZIEMER. Weakly Differentiable Functions: Sobolev Spaces and Functions of Bounded Variation.
121 LANG. Cyclotomic Fields I and II. Combined 2nd ed.
122 REMMERT. Theory of Complex Functions. *Readings in Mathematics*
123 EBBINGHAUS/HERMES et al. Numbers. *Readings in Mathematics*
124 DUBROVIN/FOMENKO/NOVIKOV. Modern Geometry—Methods and Applications. Part III.
125 BERENSTEIN/GAY. Complex Variables: An Introduction.
126 BOREL. Linear Algebraic Groups. 2nd ed.
127 MASSEY. A Basic Course in Algebraic Topology.
128 RAUCH. Partial Differential Equations.
129 FULTON/HARRIS. Representation Theory: A First Course. *Readings in Mathematics*
130 DODSON/POSTON. Tensor Geometry.
131 LAM. A First Course in Noncommutative Rings.
132 BEARDON. Iteration of Rational Functions.
133 HARRIS. Algebraic Geometry: A First Course.
134 ROMAN. Coding and Information Theory.
135 ROMAN. Advanced Linear Algebra.
136 ADKINS/WEINTRAUB. Algebra: An Approach via Module Theory.
137 AXLER/BOURDON/RAMEY. Harmonic Function Theory.
138 COHEN. A Course in Computational Algebraic Number Theory.
139 BREDON. Topology and Geometry.
140 AUBIN. Optima and Equilibria. An Introduction to Nonlinear Analysis.
141 BECKER/WEISPFENNING/KREDEL. Gröbner Bases. A Computational Approach to Commutative Algebra.
142 LANG. Real and Functional Analysis. 3rd ed.
143 DOOB. Measure Theory.
144 DENNIS/FARB. Noncommutative Algebra.
145 VICK. Homology Theory. An Introduction to Algebraic Topology. 2nd ed.
146 BRIDGES. Computability: A Mathematical Sketchbook.
147 ROSENBERG. Algebraic *K*-Theory and Its Applications.
148 ROTMAN. An Introduction to the Theory of Groups. 4th ed.
149 RATCLIFFE. Foundations of Hyperbolic Manifolds.
150 EISENBUD. Commutative Algebra with a View Toward Algebraic Geometry.
151 SILVERMAN. Advanced Topics in the Arithmetic of Elliptic Curves.
152 ZIEGLER. Lectures on Polytopes.
153 FULTON. Algebraic Topology: A First Course.
154 BROWN/PEARCY. An Introduction to Analysis.
155 KASSEL. Quantum Groups.
156 KECHRIS. Classical Descriptive Set Theory.
157 MALLIAVIN. Integration and Probability.
158 ROMAN. Field Theory.
159 CONWAY. Functions of One Complex Variable II.
160 LANG. Differential and Riemannian Manifolds.
161 BORWEIN/ERDÉLYI. Polynomials and Polynomial Inequalities.
162 ALPERIN/BELL. Groups and Representations.
163 DIXON/MORTIMER. Permutation Groups.
164 NATHANSON. Additive Number Theory: The Classical Bases.
165 NATHANSON. Additive Number Theory: Inverse Problems and the Geometry of Sumsets.
166 SHARPE. Differential Geometry: Cartan's Generalization of Klein's Erlangen Program.
167 MORANDI. Field and Galois Theory.
168 EWALD. Combinatorial Convexity and Algebraic Geometry.
169 BHATIA. Matrix Analysis.
170 BREDON. Sheaf Theory. 2nd ed.
171 PETERSEN. Riemannian Geometry.
172 REMMERT. Classical Topics in Complex Function Theory.
173 DIESTEL. Graph Theory.
174 BRIDGES. Foundations of Real and Abstract Analysis.
175 LICKORISH. An Introduction to Knot Theory.
176 LEE. Riemannian Manifolds.
177 NEWMAN. Analytic Number Theory.
178 CLARKE/LEDYAEV/STERN/WOLENSKI. Nonsmooth Analysis and Control Theory.